How the Scientis

Teaching Middle School Science Lab Procedures

by Theodore S. Abbgy

illustrated by Gary Mohrman

Senior Editor: Kristin Eclov
Editors: Cindy Barden and Christine Hood
Inside and Cover Design: Good Neighbor Press, Inc.

Good Apple
A division of Frank Schaffer Publications
23740 Hawthorne Blvd.
Torrance, CA 90505

GA1692
ISBN 0-76820106-3

Table of Contents

Chapter 3 Measurement

How the Scientist Works

Introduction

This book provides supplemental materials for middle-school science teachers. Most middle-school science teachers receive a curriculum involving physical science, earth science, life science, or general science. Teachers are advised to use many labs, hands-on activities, and cooperative learning in their teaching. In many districts, science teachers are also asked to teach processes like data collection, prediction, measurement, quantification, and so on.

Most of the supplemental materials available are specific to physical science, earth science, or life science. There are very few supplemental materials available to help teach the skills, concepts, and processes that students need *before* they attempt to learn the curriculum.

This book contains student activities, grading sheets, and transparency masters designed to help middle-school science teachers teach three vital areas required in every science class. These rules, activities, and tests have been used in urban, heterogeneous, and multicultural settings.

1 Laboratory Safety

This section contains lab safety rules, activities, and tests designed to make science labs in your school safe places for students to work and learn. These activities are designed to make students aware that accidents can happen if certain rules are not followed, and what students must do if accidents occur.

2 The Scientific Method

The activities in this chapter not only teach the scientific method, but actually give students hands-on experiments that reflect the scientific method and apply to all areas of science.

3 Measurement

Since the United States has not converted to the metric system and most of the instruments used in the science classroom use the metric system, students may not be proficient in using meters, liters, grams, and degrees Celsius. This chapter not only exposes students to the metric system, but also shows them how to use the units and instruments scientists use. This chapter also provides experiments and tests in which students use units and instruments for measurements in preparation for those used all year in science class.

The materials in this book are best introduced in the first few weeks of school, although they can be used throughout the year. Laboratory safety is best taught the first week of school and carried through on every lab project that follows.

The scientific method chapter and measurement chapter should be used after the laboratory safety chapter is completed. This information should also be introduced during the first few weeks of school before a physical science, earth science, life science, or general science unit is attempted.

Many of the activities can be completed in cooperative groups or individually. Most of the activities are at the teacher's discretion as to how they are to be used.

A list of supplies is included for each activity. The quantities are based on one class with 32 students divided into eight groups of four students each. Since most middle-school science teachers teach more than one science class you may have to figure on more of some of the consumable supplies.

Cooperative Learning

Working in Groups

Many activities in this book should be done in cooperative groups of three or four students. If possible, there should be no more than eight groups per class when labs are conducted. Teachers are advised to assign members of the class to groups and to have the same students work together in groups throughout the year. From time to time, teachers may decide to let students pick their own groups for one class period.

Teachers need to consider many factors when assigning students to groups. Conduct, creativity, intelligence, attitude, leadership abilities, gender, and maturity levels of students affect how well groups function together. Mixing students who are strong in one area with those who are strong in other areas can help groups function smoothly.

Student conduct and attitude are always of concern to teachers when assigning students to work in groups. Putting three or four students together who have behavioral or attitude problems can result in a group that is disruptive to the entire class.

It is helpful if at least one creative student is assigned to each group. Creative students assist the group in drawing, graphing, and poster-making activities. Students with strong analytical skills can help others when those skills are needed.

Given a choice, most students at the middle-school level tend to prefer working in same-gender groups. Teachers may want to encourage mixing boys and girls, as well as encouraging a blend of students with different racial and ethnic backgrounds.

Every group needs a strong leader. If the teacher does not assign a leader, one student usually surfaces as the group's leader. Mixing students with various skills, backgrounds, and abilities gives them an opportunity to work together in cooperative groups.

Social Skills

At this age, students need to be taught many social skills in order to work together in groups. In a lab situation in which students are moving around and talking in groups, it may

be necessary to remind them to "tone down" their voices to keep from disturbing others. Teachers can also cut down the noise level by not allowing students to talk to members of other groups during class time.

Some students have difficulty pacing themselves and managing their time. As a result, they fail to complete assigned tasks in the allotted time. One way teachers can help these students is by announcing how much time is left for the activity, especially if he or she notices that some groups appear to be falling behind.

When teachers need to catch the attention of the group, a catch word or phrase like "I need your attention," or "Stop!" can quickly remind them to pay attention.

When working in a lab situation, it would be easy for some students to sit back and watch other members of the group do all the work. All students need to actively contribute and participate in the group.

For some activities in this book, there is more than one correct answer. Part of cooperation is learning to come to a consensus on answers. Students need to learn that although they may not agree with the answer submitted, there are times when it is necessary to defer to the group opinion. This is especially true for activities that call for predictions.

It may be necessary for teachers to assign tasks to individuals to keep students from wandering away from the group during a lab session. One student from each group can be asked to gather and turn in materials for the whole group.

In most groups, students who are natural leaders will take on that role. However, that student may not always be the most efficient and effective leader. In some cases the teacher may need to select the group leader. This person can help the group pace themselves to complete the assignment in the time available.

When appropriate, students can be assigned as recorder, timer, or reporter. This is a good opportunity for teachers to select students who may be unwilling or shy about participating.

One person in each group should be assigned the responsibility for writing answers, recording data, and other writing for the group.

When lab experiments need to be timed, one student needs to fill the role of time-keeper for the group.

When group information needs to be reported to the class, one student can be the reporter for the group.

By helping students learn to work cooperatively in a science lab, science teachers are teaching more than science. Students also learn valuable social skills needed to be successful both in and out of the classroom.

Cross-Curricular Learning

Science teachers often work closely with other teachers, especially in the areas of safety, the scientific method, and measurement. The following suggestions provide opportunities for cross-curricular learning in these areas.

Safety

Since the unit on lab safety will be taught at the beginning of the year, it can be tied into other safety units taught in school, like bus safety, fire safety, and bike safety. The science teacher may want to coordinate safety units with other teachers. For example, he or she could contact the health teacher to suggest a unit on safety in the home or at school be taught at the same time students are learning about lab safety. Safety issues are also discussed in physical education, cooking, sewing, industrial arts, and technological education. Units on career education may focus on safety. In math, safety can be incorporated into the curriculum by making home, school, traffic, and public safety statistics available for students to analyze.

English classes provide an excellent forum for teaching students to analyze articles on safety. The articles in this book can be shared with English teachers, giving students opportunities to summarize articles and write their own essays on safety.

The Scientific Method

In this area, the scientific method relates most closely to math. Students will be collecting data as they test their hypotheses in science. That data needs to be interpreted and communicated. Much data can be communicated through various types of graphs and charts. While students are learning the scientific method in science class, math teachers can be teaching data collection and communication of data through graphs and charts.

Searching for patterns can be used in almost every subject area. The ability to recognize patterns in other areas can help students find patterns in data collected in the science lab.

Measurement

Measurement is used in many subject areas across the curriculum. Many math teachers devote entire units to measurement, teaching students to convert standard units of measure from one form to another and to convert between standard and metric units. Learning to multiply and divide decimals by units of 10, 100, and 1,000 will assist students working with the metric system in science.

Students learn in geography that most other countries use the metric system. They could study the historical background of the metric system and systems of measurement used by other cultures, both past and present. A writing assignment could include comparing and contrasting two different systems of measurement. Even in physical education, measurements are frequently used in sports and games. Swimming, track, and field events usually use the metric system of measurement.

Multiple Intelligences

This book can be applied to multiple intelligence theory. The theory simply states that there are seven intelligences, not just one. We as science teachers believe that all students are logical/mathematical thinkers. This, however, is only one of the seven intelligences. The seven intelligences include:

1. Verbal/Linguistic
2. Logical/Mathematical
3. Visual/Spatial
4. Body/Kinesthetic
5. Musical/Rhythmic
6. Interpersonal
7. Intrapersonal

There are many multiple intelligence surveys available today. Students in your class can be surveyed to see which intelligences are most prominent and which intelligences should be stressed. This book heavily stresses the logical/mathematical intelligence, however, many of the activities can be applied to the other intelligences. Suggestions for methods to teach the multiple intelligences and ideas for extension activities are included in the teacher guides.

Assessment and Evaluation

Most of the pages in this book are assessment tools of one type or another. Any assignment can be used as an assessment tool whether it is a worksheet, a lab sheet, a lab write-up, study questions, a quiz, a test, a homework assignment, a summary sheet, or a study guide.

Quizzes are provided on pages 18, 19, 41, and 76. Tests are provided on pages 24-26, 54-56, and 85.

Grading sheets can be found on pages 23, 39, 81, and 83.

References:

Computer Clip Art

Print Artist for Windows by Maxis, 1994.

(Grolier's Multimedia Encyclopedia, Release 6, 1993.)

Multiple Intelligences

Gardner, Howard. *Multiple Intelligences: The Theory in Practice.* Basic Books, 1993.

Bean Plant and Goldfish Experiments

Otto, J. H., Towle, A., and Crider, E. H. *Biology Investigations.* Holt, Rinehart and Winston, Inc., 1969.

Chapter 1
Laboratory Safety

Teacher's Guide to Laboratory Safety Rules

Laboratory Safety Rules
(pages 11 and 12 Transparency Masters)

These pages can be used as an overhead master, reproducible sheet for every student, or both. Each rule should be gone over and students should be given examples and demonstrations of these rules. There is also a section provided for rules specific to your school.

Lab Safety Agreement
(page 13)

This page is designed to communicate with parents about lab safety. Teachers, students, and parents need to know that the middle-school science lab is a safe place to work and learn. It is recommended that this page be given to every student, discussed, taken home to be signed, and returned. Keep the bottom portion of the signed agreement on file for future reference. Students should be required to have this agreement signed before they can participate in any experiments in the lab.

Lab Safety Rules Pictures
(page 14)

The lab safety rules pictures sheet provides an excellent opportunity for students to review the lab safety rules after you have discussed and demonstrated each one. This page can be used as an individual assignment, homework assignment, cooperative learning assignment, or a discussion sheet. Allow students to use the copies of the Laboratory Safety Rules to identify the pictures. Stress the fact that more than one rule may apply to each picture and that students need only choose one rule per picture.

Safe School Labs Stories
(page 15)

The two articles presented on this page are intended to provide students with situations that involve threats to lab safety. Have students take turns reading, read the stories to themselves, or read and discuss the stories in cooperative groups. For a follow-up activity, students could be asked to write stories demonstrating other lab safety situations. Be careful with this type of activity; students can get very graphic.

Safe School Labs Study Questions (pages 16 and 17)

These sheets can be reproduced and given to students to complete individually, at home, or in cooperative groups. These sheets provide follow-up for the stories on page 15. Stress that students must submit complete answers. Give examples of complete answers so students know what to do.

Safe School Labs Stories Quiz (pages 18 and 19)

This quiz has two objectives. The first objective is to make sure students understood the articles. The second objective is to prepare students for the Laboratory Safety Test on pages 24-26. It is recommended that students complete this test individually in class. This test can be checked in class to provide a formum for discussing the articles.

Lab Safety Group Activity (pages 20-22)

One objective of this group activity is to have students evaluate each picture as showing safety problems, safe practices, or both. To complete page 20 teachers must provide 10 photos, drawings, posters, or textbook photographs for students to evaluate. Pages 21 and 22 provide 10 illustrations that can be used. Allow students to work in cooperative groups to complete page 20. Accept any answer that students can justify. Another objective of this group activity is for students to see examples of lab safety rules in action.

Lab Safety Poster (page 23)

Supplies Needed

8.5" x 11" (21 cm x 27.5 cm) paper (32 sheets)
Rulers (32)
Scissors (32)
Highlighters (8 sets, various colors)
Colored pencils or markers (8 sets)
Poster board or large paper (32 sheets)
Copies of Lab Safety Rules (32)

Have students construct a lab safety poster that shows one or more of the rules in a specific category listed on pages 11 and 12. The categories are also listed in the directions on page 23. Give each student a copy of page 23. Provide students with a variety of art supplies.

Students can work individually, in cooperative groups, or complete this activity as homework. Ask students to construct a rough draft and have it approved before starting on the final copy. Since assignments like this are difficult to grade accurately, a grading sheet is provided on the bottom of the page.

Laboratory Safety Test (pages 24-26)

This 20-question test is designed as a formal check to see if students have learned their laboratory safety rules. Each student should pass the test individually in class before being allowed to perform any experiments in the lab. Before the test is given, determine the number of questions (14, 15, or 16 is recommended) that students must answer correctly before they can perform any experiments in the lab. Let students know this in advance. Students who do not correctly complete the number of designated questions must retake the test until they do so.

It is recommended that the following requirements be completed before a student can perform any experiments in the lab.

1. Have the lab safety agreement on page 13 completed and signed by a parent.
2. Correctly complete the predetermined number of questions on the Laboratory Safety Test on pages 24-26.

It is recommended that the teacher keep accurate records of which students have completed the agreement and lab safety test. Obtain copies of the agreement and lab safety test from new students who enter your class later in the year.

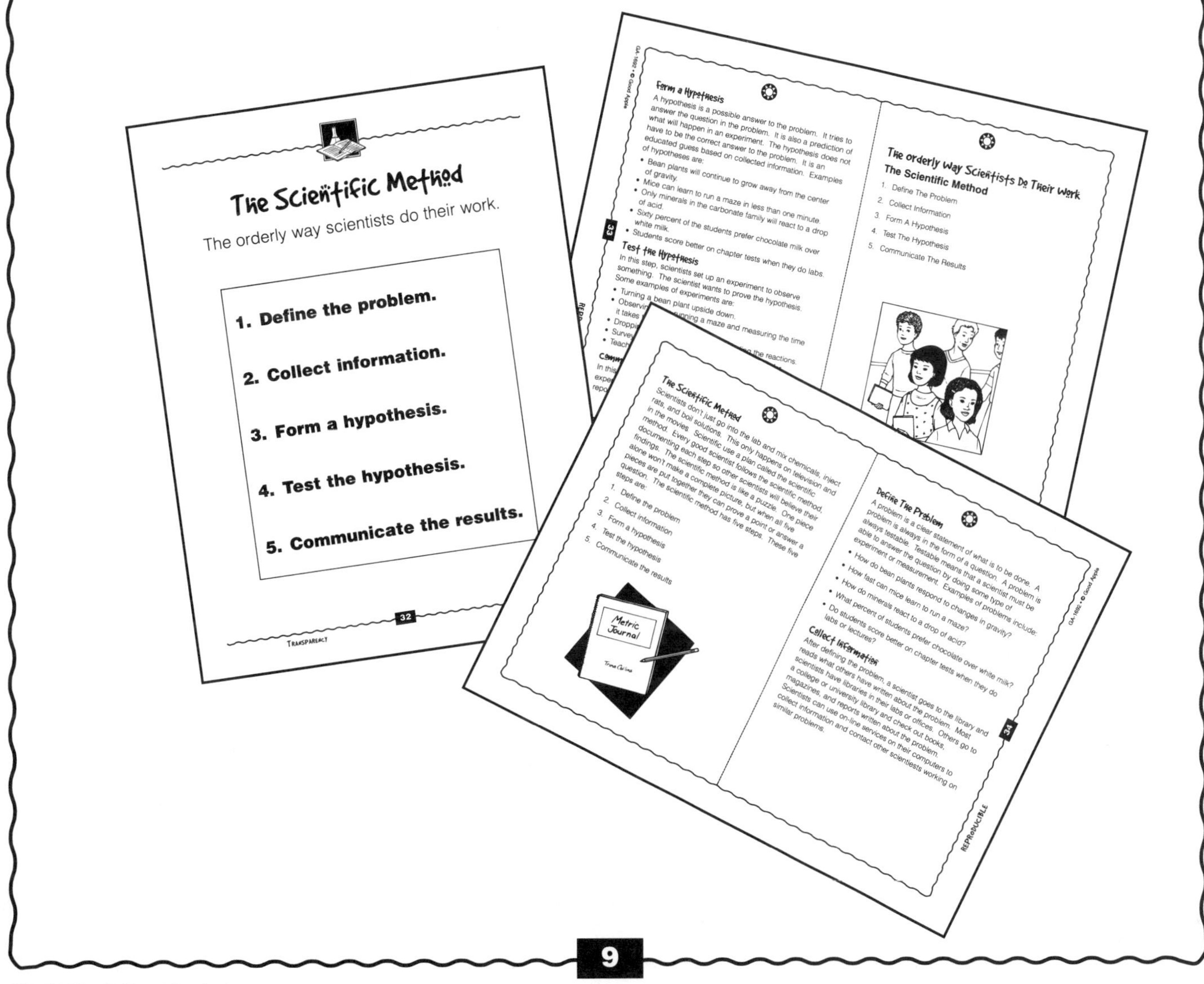

Multiple Intelligences Applications

Verbal/Linguistic–In this chapter students can explain which lab safety rules apply to the pictures and stories. Group activities allow students to discuss what they have learned with their group. After completing the poster activity, they can explain their posters to the class or group. When going over the quiz and test, students can explain their answers.

Logical/Mathematical–Students apply logical/mathematical skills when completing the activities in this chapter. The poster activity gives students an opportunity to show the rules being followed. The questions on the quiz and test allow students to apply accepted laboratory safety rules.

Visual/Spatial–Students can draw their own pictures to illustrate safety rules and what happened in the stories. The poster activity gives students an opportunity to use creativity in their designs. Students could draw a map or diagram of the lab showing where the safety supplies are located.

Body/Kinesthetic–Students can demonstrate safety rules and act out the stories to show how the accidents could have been prevented. They can act out several of the situations in the Lab Safety Group Activity or the Lab Safety Poster activity.

Musical/Rhythmic–Students can make up songs or raps that teach lab safety rules or ones to accompany their posters, creating a multi-media presentation. It is also fun for students to record their compositions on an audio cassette, make a musical video that acts out lab safety rules, or perform their compositions for the class. Students can use a tune from a well-known song or they can compose their own music.

Interpersonal–Students can rewrite the stories so that the characters act in a safe manner. They can explain the situations on their posters to the group.

Intrapersonal–Students can make lists of what they would do in reaction to safety rules and offer suggestions of how they would correct the problems found in the illustrations. They can explain what they would do if they were one of the characters in the stories. It's usually fun for them to explain what they would do if they were the teacher.

Laboratory Safety Rules

The science lab is a wonderful place to work and learn. The science lab should also be one of the safest places in the school. Oftentimes we concentrate on the lab activity itself or our friends in the class and we forget about safety. We should condition ourselves to be not only conscious of safety, but to make it the major concern while we are doing an experiment. To condition ourselves we need to practice our lab safety rules over and over again. We should study them, keep them with us when we are doing a lab, and constantly remind ourselves and others of our laboratory safety rules.

1. Prepare Properly

A. Tie back long hair and confine loose clothing.

B. Wear aprons and safety glasses when told to do so.

C. Remove long necklaces and other items like scarves or ties before starting a lab project.

D. Keep all unnecessary items off the lab table.

E. Know where all the safety and first-aid items are located in the laboratory. These items include a fire extinguisher, fire blanket, first-aid kit, safety shower, eye wash bottle, and so on.

F. Listen carefully and follow the teacher's directions.

G. Begin only when told to do so.

2. Be Careful

A. Never reach across an open flame.

B. Never put anything in or near your mouth or eyes in the science lab.

C. Never engage in horseplay in the lab.

D. Never breathe in fumes directly.

E. Never force glass because it will break.

F. Never look directly into bright lights.

G. Never leave chemical bottles open. Always replace the tops.

H. Never pour water into acid.

3. Be Considerate of Others

A. Remember the lab is a place to work and learn.

B. Be careful not to bump other students and their equipment in the lab.

C. Talk quietly in the lab and only to other members of your group.

D. Attend to the teacher quickly.

E. Work at a steady and constant pace.

F. Stay on task.

G. Treat classmates with respect.

H. Participate in lab activities.

4. Report Unusual Occurrences to Your Teacher

A. Report any equipment that is broken or not working properly.

B. Report any chemical spills immediately.

C. Report accidents immediately.

5. Clean Up

A. Dispose of all used chemicals and matches as your teacher instructs.

B. Leave your area neat and clean.

C. Return all equipment to the proper place.

D. Wash your hands.

6. Specific Lab Safety Rule for Our Lab

A. ______________________________

____

B. ______________________________

____

C. ______________________________

____

Lab Safety Agreement

The science lab is a wonderful place to work and learn. The science lab should also be one of the safest places in the school. To make the science lab the safest place in the school I agree to follow these rules:

1. I will prepare properly for lab.
2. I will be careful.
3. I will be considerate of others.
4. I will report unusual occurrences to my teacher.
5. I will clean up my area and help clean the entire lab.
6. I will follow the lab rules I have learned.

I also agree to conduct myself in a responsible manner and assume responsibility for my safety and the safety of others.

If I conduct myself in a responsible manner our school science lab will be the safest place in the school.

✂- -

Cut off and return to your science teacher.

I agree to the above conditions.

Student Signature____ Date____________

Parent Signature____

Parent Comments____

____

____

____

____

Lab Safety Rules Pictures

Name____________________________________

Date ____________________ Hour____________

Directions: Next to each picture write one lab safety rule that applies to the picture and one reason for that lab safety rule.

BANDAGES

CANDY BAR

Safe School Labs Stories

New Sweatshirt

Tom Anderson saved up his money from cutting grass and raking leaves. He had his eye on a Florida Marlins sweatshirt. It was so cool. The trouble was that it costs $42.00. When Tom had enough money he went to the mall and bought the sweatshirt.

Tom was so anxious to wear the new sweatshirt, he wore it to school the next day. He forgot that Mr. Washington, his science teacher, was conducting a science lab that day.

The students were staining onion tissue with iodine and observing the cells under the microscope. Tom was worried. He remembered that Mr. Washington said iodine will stain clothing.

Tom carefully placed drops of iodine on the onion tissue. He spilled no iodine and closed the cover as soon as he finished.

As Tom was drawing the onion cells on his lab sheet, his lab partner, Lafonzo, was making his own onion slide. Lafonzo set the open bottle of iodine on the lab table. While cleaning his microscope slide, Lafonzo accidentally knocked over the open iodine bottle. The iodine spilled, destroying the paper Tom was working on. The iodine also stained his new Florida Marlins sweatshirt. Tom was very upset. His new sweatshirt was stained forever.

Ms. Reitman's Lesson

Mrs. Reitman was the best science teacher in the school. Her labs were fun and students also learned a lot in her class.

Today her lesson was about inserting a glass tube into a rubber stopper. Tomorrow the class planned to evaporate sea water, collect the water vapor, and make fresh water. In order to conduct this experiment, the students needed to learn how to insert a glass tube into a rubber stopper.

Ms. Reitman carefully explained and demonstrated how to insert the glass tube. Then she passed out rubber stoppers and glass tubes and had the students in the class practice.

Marla couldn't get her tube into the stopper. She looked around the class. All the other students had completed the activity. Marla got frustrated and forced the glass. It broke and a sharp piece of glass cut her hand. Ms. Reitman sent Marla to the nurse.

Safe School Labs Study Questions "New Sweatshirt"

Name______________________________

Date ____________________ Hour__________

Directions: Read the "New Sweatshirt" story and answer the following questions. Use complete sentences.

1. Briefly describe the laboratory experiments that Tom and Lafonzo were trying to do.

____

____

____

2. List two lab safety rules that were ignored.

____

____

3. What could Tom have done to avoid getting his sweatshirt stained?

____

____

4. What could Lafonzo have done to avoid staining the sweatshirt?

____

____

5. What could Mr. Washington have done to avoid the accident?

____

____

6. What would be a better title for this story?

____

____

7. Would you spend $42.00 on a sweatshirt, then wear it to a science lab? Explain.

____

____

8. Was this all Lafonzo's fault? Why?______________________________

____

____

Safe School Labs
Study Questions
"Ms. Reitman's Lesson"

Name____

Date ______________________ Hour_______________

Directions: Read the "Ms. Reitman's Lesson" story and answer the following questions. Use complete sentences.

1. Briefly describe the experiment Ms. Reitman's class will do tomorrow.

____

____

____

2. List two lab safety rules that were not followed in this story.

____

____

3. What could Marla have done to avoid getting cut?

____

____

4. What could Ms. Reitman have done to avoid Marla's accident?

____

____

5. What could the other members of Marla's class have done to avoid the accident?

____

____

6. What would be a better title for this story?

____

7. Should you listen to the teacher when she or he is teaching a lesson? Why?

____

____

____

8. Who's fault was the accident? Explain.

____

____

Safe School Labs Stories Quiz

Name____________________

Date ______________ Hour__________

Directions: Read the two stories "New Sweatshirt" and "Ms. Rietman's Lesson". Write the answers to the questions in the spaces provided.

Part I Matching: Write the letter of the best answer in the space provided. Match the choices to the statements. Use each letter once.

	Statements		Choices
_____	1. A chemical used to stain onion cells.	A.	$42.00
_____	2. An instrument used to see cells.	B.	Evaporate
_____	3. Done to make fresh water from sea water.	C.	Iodine
_____	4. Student who cut herself on a glass tube.	D.	Lafonzo
_____	5. Student who spilled a bottle of iodine.	E.	Microscope
_____	6. Student who bought a new sweatshirt.	F.	Marla
_____	7. The cost of a Florida Marlins sweatshirt.	G.	Mr. Washington
_____	8. Tom and Lafonzo's science teacher.	H.	Ms. Reitman
_____	9. Marla's cool science teacher.	I.	Tom

Part II Identify the Lab Safety Rule: In the stories, several lab safety rules were broken. Put a checkmark in the space provided for the rules that were broken in the stories. If the rule was not broken in the stories, leave the space blank.

_____ 10. Prepare properly.

_____ 11. Wear safety glasses and aprons when told to do so.

_____ 12. Never engage in horseplay in the lab.

_____ 13. Never leave chemical bottles open. Replace the tops.

_____ 14. Stay on task.

_____ 15. Report any chemical spills immediately.

_____ 16. Keep all unnecessary items off the lab table.

_____ 17. Listen carefully to the teacher's directions.

_____ 18. Never force glass because it will break.

_____ 19. Never look directly into bright lights.

_____ 20. Participate in lab activities.

Safe School Labs Stories Quiz Page 2

Name________________________________

Date ____________________ Hour____________

Part III Sequencing for the "New Sweatshirt" Story: Below is a list of events that happened in the New Sweatshirt Story. Put them in order by placing 1-8 in the spaces provided to show the sequence of events.

_____ 21. Tom stains his onion cells.

_____ 22. Tom gets a stain on his new sweatshirt.

_____ 23. Tom buys a Florida Marlins sweatshirt.

_____ 24. Lafonzo leaves the top off the iodine bottle.

_____ 25. Tom rakes leaves and cuts grass to earn money.

_____ 26. Lafonzo spills the open iodine bottle.

_____ 27. Tom wears his new sweatshirt to school.

_____ 28. Tom starts to draw his onion cells.

Part IV Sequencing for the "Ms. Reitman's Lesson" Story: Below is a list of events that happened in the Ms. Reitman's Lesson story. Put them in order by placing 1-9 in the spaces provided to show the sequence of events.

_____ 29. The class practices inserting glass tubes into rubber stoppers.

_____ 30. Marla breaks a glass tube.

_____ 31. The class does a lab in which they make fresh water from sea water.

_____ 32. Marla could not get her glass tube into the rubber stopper.

_____ 33. Marla cuts her hand.

_____ 34. Ms. Reitman passes out rubber stoppers and glass tubes.

_____ 35. Ms. Reitman teaches her class the proper way to insert a glass tube into a rubber stopper.

_____ 36. Ms. Reitman sends Marla to the nurse.

_____ 37. Marla forces the glass tube into the rubber stopper.

Group # _______ Date ___________________ Names _____________________ _____________________

_____________________ _____________________

Lab Safety Group Activity

Purpose: To study laboratory pictures and evaluate safety practices.

Procedure:

1. Study the Lab Safety Group Activity Pictures.
2. Fill in the chart with your group.

Photo #	Safety Problems in the Picture	Safe Practices in the Picture
1		
2		
3		
4		
5		
6		
7		
8		
9		
10		

Lab Safety Group Activity Pictures

Lab Safety Group Activity Pictures

Lab Safety Poster and Grading Sheet

Name____________________

Date ____________________ Hour____________

1. Construct a lab safety poster showing one or more lab safety rules.
2. Design a rough draft on an 8.5" x 11" piece of paper.
3. Have the rough draft approved by your teacher.
4. Draw your poster on poster board or large paper.
5. Stick to one of the following categories.

Prepare properly | *Be careful*
Be considerate of others | *Report unusual occurrences*
Clean up | *Specific lab safety rules*

1. Completeness 0 1 2 3 4 5 6 7 8 9 10
Was the poster completed in the time allowed?

2. Neatness 0 1 2 3 4 5 6 7 8 9 10
Were the lettering, drawings, captions, and charts presented neatly?

3. Planning 0 1 2 3 4 5 6 7 8 9 10
Was there evidence that the poster was planned in advance?

4. Lesson 0 1 2 3 4 5 6 7 8 9 10
Did the poster teach a lab safety lesson?

5. Use of Color 0 1 2 3 4 5 6 7 8 9 10
Was color used to help teach the lesson and add interest to the poster?

6. Labels 0 1 2 3 4 5 6 7 8 9 10
Were all pictures, charts, diagrams, and figures labeled properly?

7. Captions 0 1 2 3 4 5 6 7 8 9 10
Were captions clearly stated and used to describe the graphics?

8. Clarity 0 1 2 3 4 5 6 7 8 9 10
Were the ideas in the poster clearly stated?
Was there an appropriate title?

9. Graphics 0 1 2 3 4 5 6 7 8 9 10
Did the graphics clearly support the lesson taught in the poster?

10. Appearance 0 1 2 3 4 5 6 7 8 9 10
Does the poster look good?

TOTAL POINTS ____________

LETTER GRADE ____________

Laboratory Safety Test

Name______________________________

Date ____________________ Hour____________

Directions: Write the best answer in the space provided.

Part I Multiple Choice: Write the letter of the best answer on the line.

_____ 1. Safety glasses or goggles are worn in the lab when the teacher instructs because
- A. They avoid eye strain
- B. They look cool
- C. They protect our eyes
- D. They help us see better

_____ 2. The lab should be cleaned up by:
- A. All students
- B. One student who had a bad day
- C. The custodian
- D. The teacher

_____ 3. A chemical has been spilled. What is the first thing you should do?
- A. Get a paper towel and clean it up.
- B. Blame your lab partner.
- C. Leave it there for the teacher to clean up.
- D. Tell the teacher and follow his or her directions.

_____ 4. Students who have long hair should:
- A. Get it cut
- B. Tie it back
- C. Wear a hat
- D. Do nothing; long hair is cool

_____ 5. What is the proper way to smell a chemical in the lab?
- A. Never smell a chemical in the lab.
- B. Wave your fingers over the chemical and inhale gently.
- C. Put the top on the bottle, wait a few minutes, and inhale gently.
- D. Put your nose on the end of the bottle and inhale deeply.

_____ 6. Items like fire extinguishers, first-aid kits, eye wash bottles, safety showers, and fire blankets are to be used by:
- A. Students
- B. Custodians
- C. Teachers
- D. Anyone who needs them

_____ 7. You are using a balance that does not work properly; what is the best thing to do?
- A. Tell the teacher
- B. Guess the mass of the object
- C. Use it anyway
- D. Try to fix it yourself

_____ 8. What is the first thing you should do when you finish using a chemical?
- A. Tell the teacher
- B. Put it in the chemical cabinet
- C. Replace the top
- D. Ask if anybody else needs it

_____ 9. What is the most important reason for having lab?
- A. To work and learn
- B. To have fun
- C. To talk to your group members
- D. To make science easier

Laboratory Safety Test Page 2

Name______________________________

Date ____________________ Hour______________

______ 10. Used chemicals should be:

A. Thrown in the waste can
B. Disposed of according to the teacher's directions
C. Saved for the next class
D. Rinsed down the drain

______ 11. You are doing an experiment in the lab. Suddenly you remember you have a stick of gum in your book bag. What should you do?

A. Wash your hands, then put the stick of gum in your mouth.
B. Throw the gum away.
C. Chew the gum when the teacher isn't looking.
D. Wait until after school.

______ 12. You picked up a dirty test tube with a sticky substance on the outside. You get some of this stuff on your hand. What is the first thing you should do?

A. Wipe your hand on your apron.
B. Wash your hands and tell the teacher.
C. Wipe your hands on the guy next to you.
D. Try to figure out what the sticky stuff is.

Part II Lab Safety Story: Read the story carefully. Then answer the multiple-choice questions 13-20 on page 41. Write the letter of the best answer on your answer sheet.

Crystal Clear

The students in Ms. Harper's science class were making crystals of copper sulfate, a poisonous chemical. The day before the lab, Ms. Harper explained how to make crystals. She demonstrated dissolving as much copper sulfate powder in 100 mL of water as possible, even if this meant heating the copper sulfate solution to boiling. When no more copper sulfate would dissolve, students were taught to shut off the heat on the hot plate, let it cool, and hang a seed crystal in the solution. She then told the students to let it sit for a few days.

On the day of the lab Ms. Harper demonstrated, once again how to make crystals. While Ms. Harper demonstrated, one lab group did not pay attention. Tasha colored in the pictures on the lab sheet. Michelle thought about the cute guy in her math class. Annisha wrote a note to her friend in another class. When Ms. Harper completed her demonstration, Tasha, Michelle, and Annisha had no idea what they should do.

When the lab started, Tasha tried to read the directions on the lab sheet. Michelle filled her 600-mL beaker with water, poured a pile of copper sulfate into the water, overflowed the beaker, and wiped up the mess with a paper towel. Annisha finished her note. Then she turned on the hot plate and reached over it to get the overflowing beaker to heat.

Three days later that group was the only group without crystals. The girls wondered why they received an "F" on the lab.

Laboratory Safety Test Page 3

Name____

Date ____________________________ Hour________________

______ 13. All of the following rules were broken by this group except:

A. Keep unnecessary items off the lab table.
B. Listen to the teacher's directions.
C. Begin only when told to do so.
D. Attend to the teacher quickly.

______ 14. What should the girls in this group have done when the lab started and they didn't know what to do?

A. Pretend they knew what they were doing.
B. Ask the members of another group what to do.
C. Copy what the members of another group were doing.
D. Read the directions carefully and ask Ms. Harper.

______ 15. Why should Michelle have gone to Ms. Harper after she overflowed the beaker?

A. To ask about the proper way to clean up a copper sulfate spill.
B. To blame Tasha and Annisha for the spill.
C. To ask for a pass to the nurse.
D. To distract her while Tasha and Annisha cleaned up the spill.

______ 16. What was the best thing the girls could have done to avoid getting an "F" on the lab?

A. Ask Ms. Harper for extra credit assignments.
B. Do the experiment over at home.
C. Listen both times Ms. Harper demonstrated the experiment.
D. Go to the library and look up crystals.

______ 17. What can Ms. Harper do to help the students in this group avoid future failures?

A. Teach the girls to pay attention during explanations.
B. Move the girls to different groups.
C. Suspend the girls from class for three days.
D. Make the girls come in after school and clean up the lab.

______ 18. Failing a lab is bad, but not the worst thing that can happen when lab safety rules are not followed. What is the worst thing that can happen?

A. More homework
B. A bad report card
C. Suspension from class
D. Student gets injured

______ 19. Which student did the best thing when the girls didn't know how to start the lab?

A. Annisha
B. Michelle
C. Tasha

______ 20. What would be the best title for this story?

A. A Group of Failures
B. Listen to the Teacher
C. Students Injured in Lab Fire
D. Writing Notes in Class

Chapter 2
The Scientific Method

Teacher's Guide to the Scientific Method

The Scientific Method Transparency Master (page 32)

Use this transparency master to introduce and explain the scientific method. Make sure students understand that the scientific method cannot always be followed step by step, although scientists try to do so. Stress the fact that scientists often rely on reasoning and imagination. Explain that the problem is always stated in the form of a question. Let students know that scientists collect much of their information by reading what others have written and done. Define the hypothesis as a possible answer to the problem. Experiment means to test a hypothesis. Help students understand that scientists must write reports on what they have done to communicate their results.

Because of TV and movies, students often have a distorted picture of what a scientist's lab looks like. They often picture colored chemicals bubbling and smoking as they pass through a maze of tubes and test animals in cages. Help them understand that labs often contain books and computers for researching and writing results.

The Scientific Method Pamphlet (pages 33 and 34)

These pages can be copied back-to-back and folded to make a pamphlet. Make a classroom set for students to use as a reference in class. These pages provide an explanation of the scientific method and an explanation and examples of each step in the scientific method.

Scientific Method Pictures (page 35)

Each picture on this sheet shows at least one step of the scientific method. Have students use the pamphlet on pages 33 and 34 to identify one step shown by the picture. After students have completed identifying the pictures, have them discuss their answers. This page provides an excellent format for reviewing the scientific method.

Scientific Method Study Questions (pages 36 and 37)

These questions are another format for reviewing the scientific method. Let students refer to the pamphlet on pages 33 and 34 to find their answers. It's a good idea to discuss the answers when students have completed pages. These pages work well in cooperative groups or for individual students.

Scientific Method Mini-Posters and Grading Sheets (pages 38 and 39)

Supplies Needed

Rulers (32)
Highlighters (8 sets, various colors)
Colored pencils or markers (8 sets)

Have students draw mini-posters in each window on page 38. The mini-posters should teach one idea on each step in the scientific method. A grading sheet has been provided on page 39. If you decide to use the grading sheet, students should see a copy of it before they attempt to draw mini-posters. This assignment works best individually, however it can also be a successful assignment in cooperative groups. Provide the students with a variety of art supplies.

Scientific Method Homework (page 40)

This sheet is specifically designed as a homework assignment. Students are asked to think up their own experiments. They can either make everything up or they can actually do the experiments. If students actually do the experiments, you may want to reward them by awarding bonus points or letting them demonstrate their experiments in class.

Scientific Method Quiz (page 41)

This 15-question quiz is intended for individual students. This quiz can be done with or without the pamphlet on pages 33 and 34. It can also be checked in class as another forum for discussing the scientific method.

Bean Plant Experiment (pages 42-47)

Supplies Needed

Pinto bean seeds (small bag)
Potting soil (large bag)
Peat pots (24 small)
Masking tape (8 rolls)
Ring stands (16)
Permanent markers (8)
Paper clips (8)
Funnels (8)

This can be a successful activity if done as a demonstration or as a student lab. This can also be a very unsuccessful activity if the bean plants do not germinate. Take the following steps to insure a high germination rate:

1. Buy fresh beans or bean seeds (pinto beans work best in labs).
2. Soak the beans for 12-24 hours before the students plant them.
3. Use potting soil rather than dirt from the school grounds.
4. Use peat pots rather than clay or plastic pots. Peat pots are easiest to hang upside down.

Once the plants have germinated and grown about 4 inches (10 cm) high, students are ready to mount two of their three plants. Mounting is best done on a dry pot with lots of masking tape. It's easy to water the mounted plants. Poke a small hole in the side or bottom of the peat pot, insert a small glass funnel, pour the water through the funnel, and then remove the funnel. Teach students this process and have them water their own plants.

Once plants B and C show a change in the direction of growth, students may draw or photograph their plants and complete the lab sheets on pages 42 and 43, the lab write-up on page 44, and the summary sheets on pages 45-47. Have students answer the summary questions at the bottom of the write-up. The experiment, write-up, and summary work best in groups.

Define a controlled experiment to students as any experiment that has many trials, many controls, and one variable. Point out the controls and variable in the bean plant experiment and give other examples. Stress the fact that not all experiments are controlled.

Measuring Air Temperature Lab (pages 48-50)

Supplies Needed

Celsius thermometers (8)
Ice (1 bag)
500-mL glass or plastic beakers (8)
Stopwatches (8)

Students will need a mini-lesson on how to read a Celsius thermometer before they attempt this lab. Caution students about safety concerns with this lab because it calls for a thermometer to be shaken. Please demonstrate safe and efficient ways to shake a thermometer. Students should also receive a mini-lesson on how to make a line graph before they attempt this experiment.

This experiment teaches two things–the steps in the scientific method and that a shaken thermometer reacts faster than a still thermometer. This experiment also provides a forum for teaching measuring temperature, collecting, and reporting data.

If keeping ice is a problem in your school buy it or make it and keep it in a cooler. If stopwatches are not available in your lab, check with the gym teacher or use an alarm clock with a second hand.

The Goldfish Experiment (pages 51-53)

Supplies Needed

Feeder goldfish (8)
Fish tank or large jar
Small fish net
Ice (1 bag)
500-mL glass beakers (8)
Stopwatches (8)

The goldfish experiment is good for teaching the scientific method. It works very well with feeder goldfish for two reasons. First, feeder goldfish are tough and can live in cold temperatures. Second, feeder goldfish cost very little in most pet stores. Any glass container that holds more than two liters of water will work well as a temporary aquarium for a dozen feeder goldfish.

If keeping ice is a problem in your school, buy it or make it and keep it in a cooler. If stopwatches are not available in your lab check with the gym teacher or use an alarm clock with a second hand.

The goldfish experiment is not only excellent for teaching the scientific method, but it also teaches data collection and graphing. Students may need a mini-lesson before this lab on how to make a line graph. They should also complete the Metric System Temperature Lab page 84 to learn how to read a thermometer.

Explain to students that the cold water temperatures do not affect the health of the fish even if their breathing does slow down. Remind them that goldfish can live in ponds that ice over in the winter.

This experiment and the following write-up is best done in cooperative groups. When the experiment is over, explain to students that the reason the fish's breathing slowed down was because colder water dissolves oxygen more efficiently than water at room temperature. The metabolism of goldfish slows down in colder water.

Scientific Method Test (pages 54-56)

This test measures what students know about the scientific method. It also presents an experiment called Maria's Mold and asks students to define the parts of the scientific method. It is best to administer this test to students individually.

Multiple Intelligences Applications

Verbal/Linguistic–Students who learn efficiently in this manner will find it helpful to discuss the activities and their answers with members of the group or class. They can explain what is happening in the mini-posters they drew. Students can record their data as they complete the experiments and demonstrate their work to the class or group. Drawing graphs provides a wonderful opportunity for students to visually present the results of their experiments. When going over the quiz and test, students can explain their answers.

Logical/Mathematical–Students apply the logical steps of the scientific method to the pictures and activities in this chapter. The questions on the quiz and test allow students to apply the scientific method. Students continue this type of learning as they complete the step-by-step procedures for the various experiments.

Visual/Spatial–Many visuals illustrating the scientific method are presented in this chapter. Students could be asked to draw a comic strip showing the scientific method related to the experiment they design in the Study Questions activity. The poster activity gives students an opportunity to use creativity in their designs. They can draw illustrations for the experiments they complete. While students observe the growth of the bean plants, they can provide visuals by making flip charts of the plant growth from day to day. They could also photograph the plants each day. Completing the graphs also provides students with an opportunity for visual/spatial learning.

Body/Kinesthetic–Students can demonstrate what is happening in the pictures for the Scientific Method Pictures activity or make up short skits to present to the class. They can demonstrate the experiment they designed in the Study Questions activity as well as the other experiments in the chapter.

Musical/Rhythmic–As an extension to what students learn about the scientific method, you could tell a story about a lab experiment involving mice and a maze. Students could make up songs or raps that take the point of view of the mouse or of the goldfish in the Goldfish Experiment activity.

Interpersonal–Activities done in groups allow students to discuss the activity with other members of the group. They can describe the experiment they designed in the Study Questions activity and in the Mini-Poster activity. When quizzes and tests are checked in class, students can discuss their answers.

Intrapersonal–When completing individual activities, students can "intrapersonalize" those that ask them what they think. Students can intrapersonalize their mini-posters by drawing themselves in the posters. Intrapersonalization also occurs when students are allowed to think up their own experiments. They can compare their own hypothesis to the results their group obtained in the experiments and to their own results. You may want to ask students how they feel when their results and hypothesis agree and how they feel when the two do not agree.

The Scientific Method

The orderly way scientists do their work.

1. Define the problem.

2. Collect information.

3. Form a hypothesis.

4. Test the hypothesis.

5. Communicate the results.

Form a Hypothesis

A hypothesis is a possible answer to the problem. It tries to answer the question in the problem. It is also a prediction of what will happen in an experiment. The hypothesis does not have to be the correct answer to the problem. It is an educated guess based on collected information. Examples of hypotheses are:

- Bean plants will continue to grow away from the center of gravity.
- Mice can learn to run a maze in less than one minute.
- Only minerals in the carbonate family will react to a drop of acid.
- Sixty percent of the students prefer chocolate milk over white milk.
- Students score better on chapter tests when they do labs.

Test the Hypothesis

In this step, scientists set up an experiment to observe something. The scientist wants to prove the hypothesis. Some examples of experiments are:

- Turning a bean plant upside down.
- Observing mice running a maze and measuring the time it takes them.
- Dropping acid on minerals and recording the reactions.
- Surveying students on what kind of milk they prefer.
- Teaching lessons using labs and lectures.

Communicate the Results

In this step a scientist writes a report on the results of the experiment. Scientists present these results by writing reports and publishing them in books and magazines.

The orderly Way Scientists Do Their Work

The Scientific Method

1. Define the Problem
2. Collect Information
3. Form a Hypothesis
4. Test the Hypothesis
5. Communicate the Results

The Scientific Method

Scientists don't just go into the lab and mix chemicals, inject rats, and boil solutions. This only happens on television and in the movies. Scientific use a plan called the scientific method. Every good scientist follows the scientific method, documenting each step so other scientists will believe their findings. The scientific method is like a puzzle. One piece alone won't make a complete picture, but when all five pieces are put together, they can prove a point or answer a question. The scientific method has five steps. These five steps are:

1. Define the problem
2. Collect information
3. Form a hypothesis
4. Test the hypothesis
5. Communicate the results

Define the Problem

A problem is a clear statement of what is to be done. A problem is always in the form of a question. A problem is always testable. *Testable* means that a scientist must be able to answer the question by doing some type of experiment or measurement. Examples of problems include:

- How do bean plants respond to changes in gravity?
- How fast can mice learn to run a maze?
- How do minerals react to a drop of acid?
- What percent of students prefer chocolate milk over white milk?
- Do students score better on chapter tests when they do labs or lectures?

Collect Information

After defining the problem, a scientist goes to the library and reads what others have written about the problem. Most scientists have libraries in their labs or offices. Others go to a college or university library and check out books, magazines, and reports written about the problem. Scientists can use on-line services on their computers to collect information and contact other scientists working on similar problems.

Scientific Method Pictures

Name________________________________

Date ____________________ Hour____________

Directions: Write one scientific method that applies to each picture.

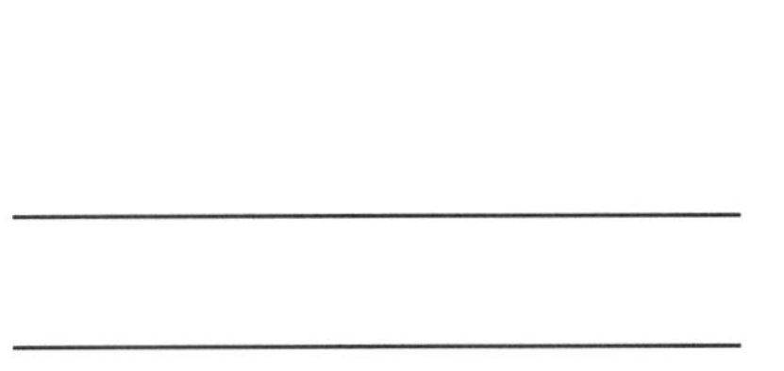

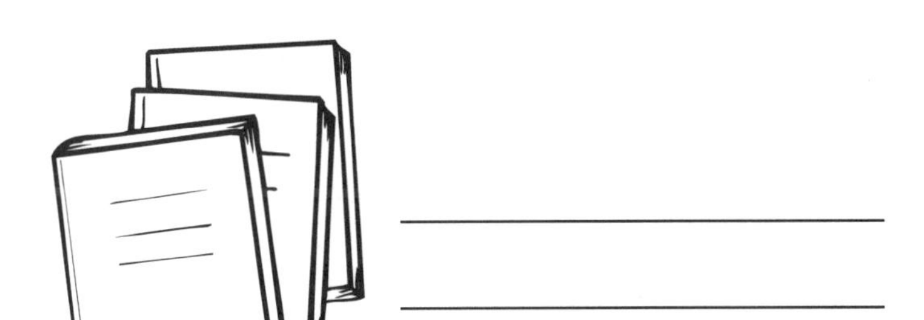

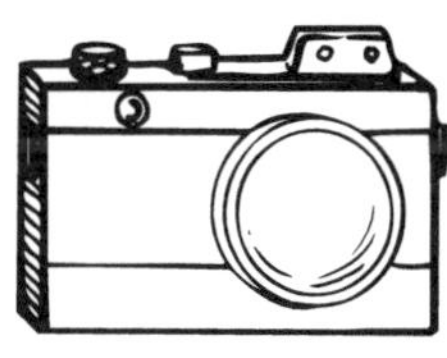

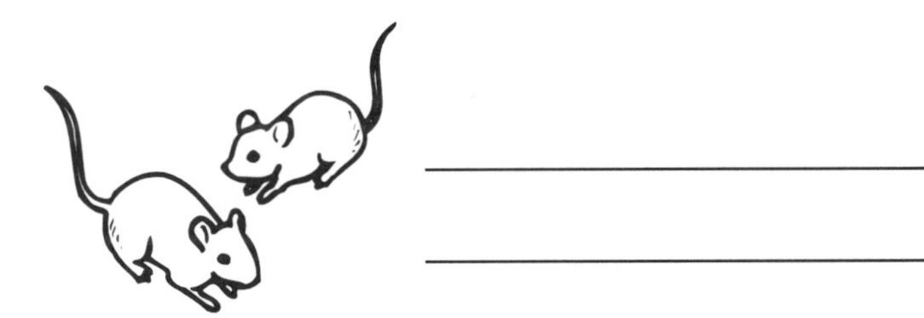

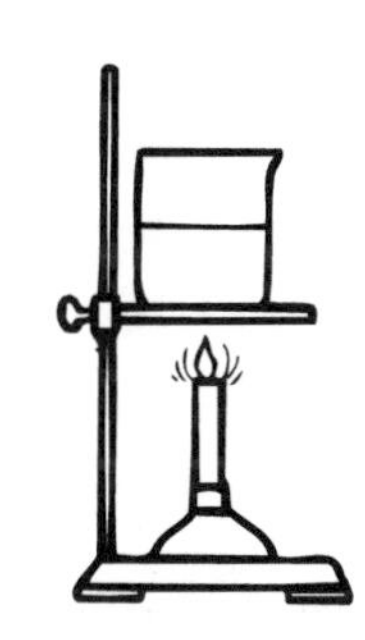

Scientific Method Study Questions

Name______________________________

Date ______________ Hour__________

Directions: Carefully read the pamphlet about the scientific method. Write short complete answers to the study questions below. Refer to the pamphlet to help you answer the questions.

1. List ten things that you think should be in a scientist's laboratory.

______________________ ______________________

______________________ ______________________

______________________ ______________________

______________________ ______________________

______________________ ______________________

2. In the pamphlet there were five examples of problems. Read the five examples, then write a new one of your own. It can be about any topic.

____

____

____

3. Describe how you would collect information about the problem you wrote in question #2. Indicate what kinds of books and articles you would have to read.

____

____

____

4. Write a hypothesis for the problem that you wrote in question #2, and collected information about in question #3.

____

____

____

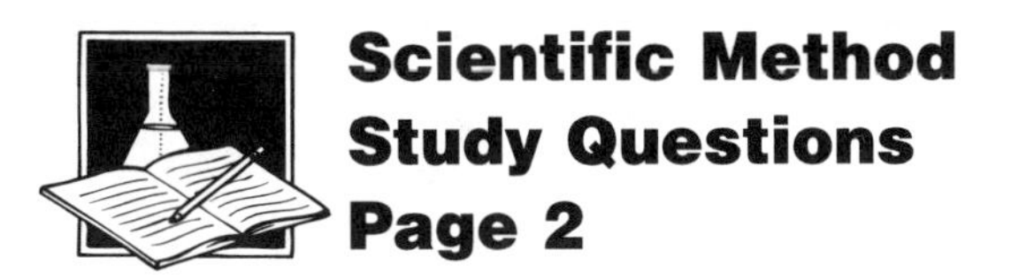

Scientific Method Study Questions Page 2

Name________________________________

Date ____________________ Hour____________

5. Describe an experiment that will prove the hypothesis you wrote in question # 4. You may make up an experiment or describe one you already know that addresses your hypothesis.

____

____

____

6. Write a report that describes the results you got in question #5. You you may make up results or describe results you already know.

____

____

____

7. Fill in the blank spaces in the model of the scientific method.

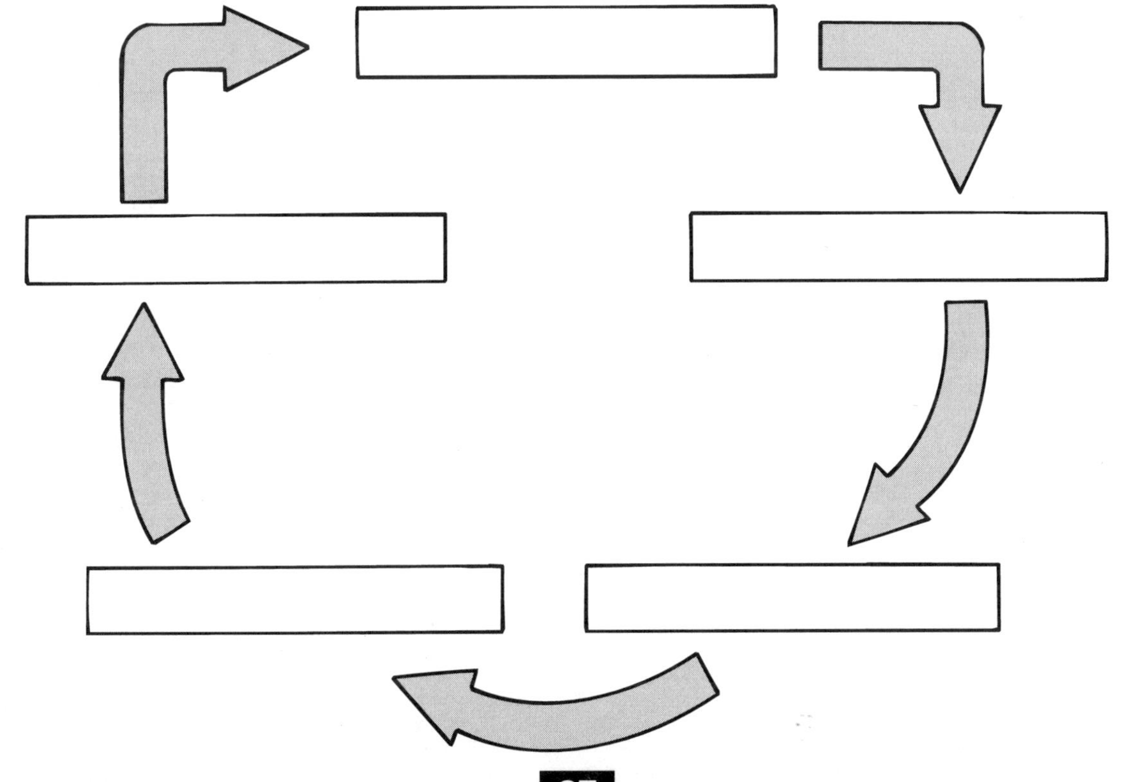

Scientific Method Mini-Posters

Name________________________________

Date ____________________ Hour____________

Directions: Draw mini-posters of each step of the scientific method in the spaces below. Each mini-poster should show neatness, completeness, planning, color, labels, captions, clarity, and graphics.

1	2	3	4	5

Scientific Method Mini-Posters Grading Sheet

Name____

Date ____________________________ Hour________________

1. Completeness Were the mini-posters completed in the time allowed?	0 1 2 3 4 5 6 7 8 9 10
2. Neatness Were the lettering, drawings, captions, and charts presented neatly?	0 1 2 3 4 5 6 7 8 9 10
3. Planning Was there evidence that the mini-posters were planned in advance?	0 1 2 3 4 5 6 7 8 9 10
4. Lesson Did the mini-posters teach a scientific method lesson?	0 1 2 3 4 5 6 7 8 9 10
5. Use of Color Was color used to help teach the lesson and add interest to the mini-posters?	0 1 2 3 4 5 6 7 8 9 10
6. Labels Were all pictures, charts, diagrams, and figures labeled properly?	0 1 2 3 4 5 6 7 8 9 10
7. Captions Were captions clearly stated and used to describe the graphics?	0 1 2 3 4 5 6 7 8 9 10
8. Clarity Were the ideas in the mini-posters clearly stated? Was there an appropriate title?	0 1 2 3 4 5 6 7 8 9 10
9. Graphics Did the graphics clearly support the lesson taught in the mini-posters?	0 1 2 3 4 5 6 7 8 9 10
10. Appearance Do the mini-posters look good?	0 1 2 3 4 5 6 7 8 9 10

TOTAL POINTS ____________

LETTER GRADE ____________

Scientific Method Homework

Name____

Date ______________________ Hour______________

Directions: Think up your own experiment. You don't actually have to do the experiment, but you can if you wish. Use the Scientific Method Pamphlet to write it up.

State the problem:____

Describe how you would collect the information: _______________________________________

State the hypothesis: ____

Describe the experiment: ____

Write a report explaining the results of your experiment: __________________________________

Scientific Method Quiz

Name_______________________________________

Date ____________________ Hour____________

Part I Matching: Match the choices to the statements. Write the letter of the best choice in the space provided. Use each letter once.

STATEMENTS

_____ 1. A plan scientists use to solve problems

_____ 2. A clear statement of what is to be done in the form of a question

_____ 3. A possible answer to the problem

_____ 4. Testing the hypothesis

_____ 5. Read books and magazines on a topic

CHOICES

A. Experiment

B. Hypothesis

C. Problem

D. Scientific method

E. Collect information

Part II Problem or Hypothesis? Write an A if the statement is a problem. Write a B if the statement is a hypothesis.

_____ 6. What percent of students prefer chocolate-chip cookies over oatmeal cookies?

_____ 7. Bees tend to prefer clover to alfalfa.

_____ 8. How does temperature affect goldfish respiration?

_____ 9. Do students do better on chapter tests if they do worksheets or hands-on activities?

_____ 10. Trees tend to be taller in forests than in the open.

Part III The Scientific Method: Match the parts of the scientific method to the parts of the mineral experiment. Use each choice once.

MINERAL EXPERIMENT

_____ 11. Place a drop of acid on a mineral sample and record the results.

_____ 12. How do minerals react to a drop of acid?

_____ 13. Get books and read about minerals and acids.

_____ 14. Write a report about how minerals react to acid.

_____ 15. Carbonate minerals will react to acid.

SCIENTIFIC METHOD

A. Define the problem

B. Collect information

C. Form a hypothesis

D. Test the hypothesis

E. Communicate the results

Group # ______ Date ________________ Names ________________ ________________

________________ ________________

Bean Plant Experiment

Purposes: To observe how bean plants react to changes in gravity. To practice the scientific method.

Procedure:

1. Use permanent marker and masking tape to label three peat pots *A, B,* and *C.*
2. Fill the three labeled peat pots almost full of potting soil.
3. Plant six bean seeds (two in each pot).
4. Water each pot.
5. Place all three pots in a sunny place.
6. Let stand for a couple of weeks until each bean plant grows about four inches (10 cm) tall.
7. Water the plants every Monday, Wednesday, and Friday.
8. Give each plant an equal amount of water and sunlight.
9. When the plants are about four inches (10 cm) tall, do the following to each plant.

PLANT A–Leave this plant as is. Do not change this plant.

PLANT B–Use masking tape to mount this plant to a ring stand as shown in Figure B.

PLANT C–Use masking tape to mount this plant to a ring stand as shown in Figure C.

Figure B

Figure C

10. Predict what will happen to each plant.

PLANT A ____

____

PLANT B ____

____

PLANT C ____

____

Group # ______ Date ________________ Names ________________ ________________

________________ ________________

Bean Plant Experiment Page 2

11. Let the plants grow for a few more days, giving them equal amounts of sunlight and water. If it is difficult to water plants B and C, pierce pots B and C with a paper clip, insert a funnel into the hole and pour the water into the funnel.
12. After step 11 has been completed, record what happened to each plant.

PLANT A ____

____

PLANT B ____

____

PLANT C ____

____

13. Cut the peat pots open and observe what happened to the roots of each plant. Record your observations.

ROOT A ____

____

ROOT B ____

____

ROOT C ____

____

14. Write a lab report based on what you observed in this experiment about bean plants and gravity. Also include in your report a brief summary of what you did.

____

____

____

____

____

Group # _______ Date ________________ Names ____________________ ____________________

____________________ ____________________

Bean Plant Experiment Write-up

Directions: Use your bean plant lab sheet to answer the following questions. Write your answers on this sheet.

1. What was the problem in the experiment? ____________________

____

2. What was your group's hypothesis? (See step 10 on your lab sheet.)

____

____

____

3. A controlled experiment has many trials, many controls, and one variable. Trials are the number of times an experiment is tried. Controls are things in the experiment that did not change. The variable is the one thing in the experiment that did change. Was the bean plant experiment a controlled experiment?

____

Why? ____________________

____

4. What is the variable in the experiment? ____________________

____

5. What are some things in the experiment that were controlled? ____________________

____

____

6. In the space to the right draw a picture of the bean plant that was controlled.

Group # ______ Date ________________ Names __________________ __________________

__________________ __________________

Bean Plant Experiment Summary Sheet

Directions: Read the following summary. Answer the 15 multiple-choice questions about the summary, the lab sheet, and the write-up.

Define the Problem

You and the members of your group probably wrote an excellent problem for this experiment in question #1 of the write-up. Please check your problem against this problem:

What is the effect of gravity on the stems and roots of bean plants?

Remember, this is only one possible wording for the problem. Yours may be worded differently, and still be correct.

Collect Information

At this point a scientist would have gone to the library and collected information about how gravity affects plants. This step in the scientific method was already done for you by your teacher.

Form a Hypothesis

In this experiment there were several possible hypotheses you could have written in step 10 on your lab sheet. These possibilities include:

1. Gravity has no effect on the roots and stems of bean plants.
2. Roots will grow away from the center of gravity and stems will grow toward the center of gravity.
3. Roots and stems will grow toward the center of gravity.
4. Roots and stems will grow away from the center of gravity.
5. Roots will grow toward the center of gravity, and stems will grow away from the center of gravity.
6. Bean plants will die if the center of gravity is changed.

Test the Hypothesis

Your hypothesis was tested by doing an experiment and observing what happened. Here three bean plants were germinated under regular conditions. *Germinated* means when the stem first begins to pop out of the potting soil. After the stems grew to a height of about four inches (10 cm) two of the three bean plants were turned and all three bean plants were allowed to grow in the same conditions. Observations were written about the stems and roots.

Communicate the Results

In step 14 of your lab sheet you were asked to communicate your results. If all went well, your results should have been the same as hypothesis #5. *Roots will grow toward the center of gravity, while stems grow away from the center of gravity.*

Group # ______ Date ________________ Names ________________ ________________

________________ ________________

Summary Sheet Page 2

Directions: Circle the letter of the best answer. Use the summary sheet, the write-up, and the lab sheet to find your answers.

Summary Questions

1. What is it called when a little stem first pops out of the soil?

   A. Control
   B. Germinate
   C. Gravity
   D. Problem

2. Which hypothesis should really have happened?

   A. #1
   B. #2
   C. #3
   D. #4
   E. #5
   F. #6

3. What part of the scientific method should be in the form of a question?

   A. Conclusions
   B. Experiment
   C. Hypothesis
   D. Problem

4. How many plants had to be planted in the experiment?

   A. one
   B. two
   C. three
   D. four

5. What is the name of the force that seems to attract the roots?

   A. Germination
   B. Gravity
   C. Hypothesis
   D. Problem

6. Which part of the scientific method tries to answer the question?

   A. Define the problem
   B. Collect information
   C. State the hypothesis
   D. Communicate the results

7. The problem must always be in the form of:

   A. An experiment
   B. A hypothesis
   C. An observation
   D. A question

8. A controlled experiment has all of the following except

   A. Many problems
   B. Many trials
   C. Many controls
   D. One variable

Group # _______ Date ____________________ Names ______________________ ______________________

______________________ ______________________

Summary Sheet Page 3

9. All of the following were controlled in the experiment except

   A. Plant A
   B. Plant B
   C. Amount of sunlight
   D. Amount of water

10. The variable in the experiment was

   A. Plant A
   B. Plants B and C
   C. Amount of sunlight
   D. Amount of water

11. Which of the following would be the best problem for the bean plant experiment?

   A. Roots will grow toward gravity and stems will grow away from gravity.
   B. How much potting soil is needed to grow bean plants?
   C. Do plants with more water grow better than plants with less water?
   D. How does the growth of stems and roots depend on gravity?

12. The part of the experiment that changes is the:

   A. Control
   B. Hypothesis
   C. Trial
   D. Variable

13. The part of the experiment that does *not* change is the:

   A. Control
   B. Hypothesis
   C. Trial
   D. Variable

14. Each time an experiment is tried it is called a:

   A. Control
   B. Hypothesis
   C. Trial
   D. Variable

15. What would be the best title for this lab?

   A. Bean Plants and Sunlight
   B. Water and Beans
   C. Water, Sunlight, and Gravity
   D. Bean Plants and Gravity

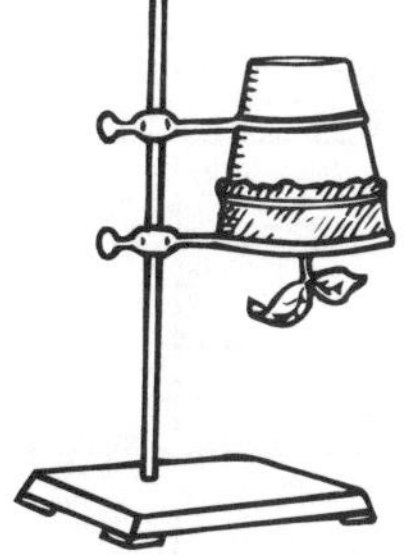

Group # ______ Date ______________ Names ______________ ______________

______________ ______________

Measuring Air Temperature Lab

Purposes: To compare the reaction time of thermometers. To study the scientific method.

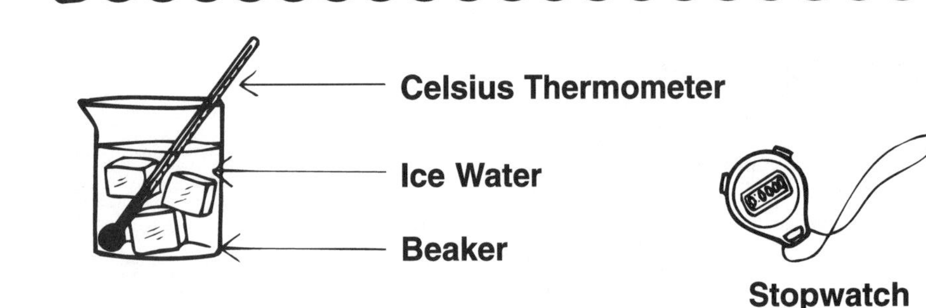

Procedure:

1. Measure and record the room temperature. ______________________
2. Make ice water in the beaker.
3. Put the thermometer into the ice water until the temperature gets close to 0 degrees Celsius.
4. Take the thermometer out of the ice water. Start the stopwatch and record the temperature in the **0 Minutes** section of Table 1.
5. Shake the thermometer and record the temperature every minute in Table 1 until the temperature reaches room temperature.
6. When Table 1 is completed, repeat steps 3-5. This time hold the thermometer still instead of shaking it. Record your data in Table 2.

Table 1–The Shaken Thermometer

Minutes	Temperature (°C)
0	
1	
2	
3	
4	
5	
6	
7	

Table 2–The Still Thermometer

Minutes	Temperature (°C)
0	
1	
2	
3	
4	
5	
6	
7	

Group # ______ Date ______________ Names ________________ ________________

________________ ________________

Measuring Air Temperature Lab Page 2

Graph: Make a line graph below for Table 1 and Table 2. Use two different colors to graph each table's data.

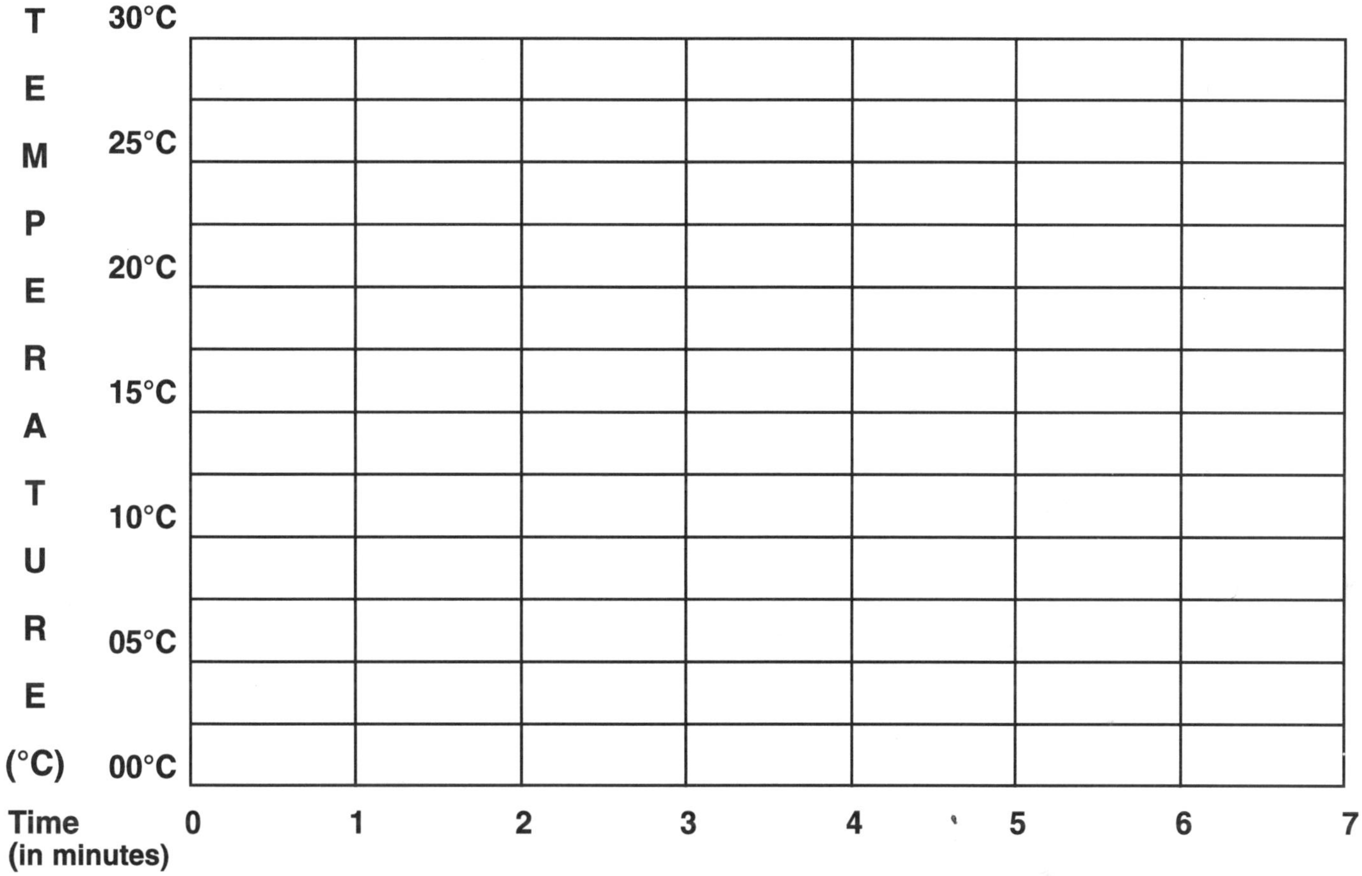

TEMPERATURE (°C)

30°C

25°C

20°C

15°C

10°C

05°C

00°C

Time (in minutes) 0 1 2 3 4 5 6 7

Directions: Give short, complete answers to the following questions.

1. Why did the thermometer in the second experiment return to room temperature more slowly than the first?

____

____

2. Would a very small thermometer react faster or slower than a larger thermometer? Why?

____

____

3. Write an appropriate problem for this experiment. ______________________

____

____

Group # ______ Date ________________ Names ________________ ________________

________________ ________________

Measuring Air Temperature Lab Page 3

4. There are three possible hypotheses for this experiment. List all three below.

5. Which hypothesis actually happened in your experiment?

6. The part of an experiment that changes is called the variable. What is the variable in this experiment?

7. Communicate your results. Write a lab report for this experiment. Your lab report should include a brief statement of the problem, a hypothesis, a summary of the experiment, and a brief conclusion you can draw from this experiment.

   Statement of the problem:

   Hypothesis:

   Summary of the experiment:

   Conclusion:

Group # _______ Date ____________________ Names ____________________ ____________________

____________________ ____________________

The Goldfish Experiment

Problem: How does water temperature affect goldfish respiration?

Part I

Hypothesis: Write your own hypothesis in the space below.

____

____

Procedure:

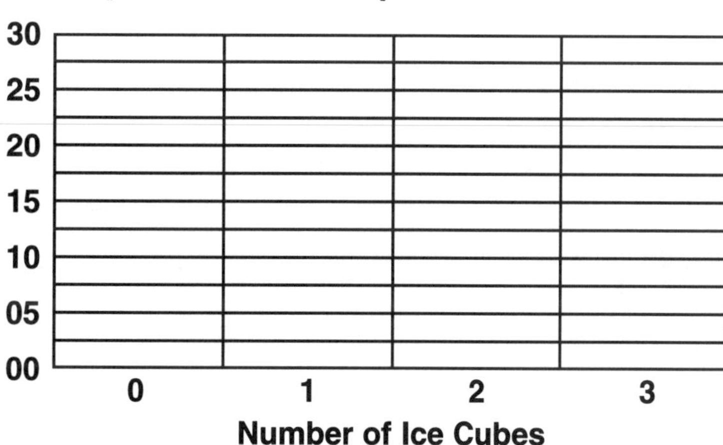

1. Place a goldfish in a beaker with enough water to keep the fish alive.
2. Give the goldfish a few minutes to get used to the water.
3. Count the respirations of the goldfish by counting the number of times the gills open and close in 15 seconds. Use a stopwatch to measure the time.
4. Multiply that number by four to get the respirations per minute.
5. Record the number of respirations per minute in the 0 section on Table 1.
6. Add an ice cube to the water and repeat steps 2-5. Repeat with two and three ice cubes. The ice cubes will not harm the fish.

Observations: Fill in Table 1, then make a line graph on Graph 1.

Table 1–Data

# of ice cubes	Goldfish respirations per minute
0	
1	
2	
3	

Graph 1–Goldfish Respirations

RESPIRATIONS

30
25
20
15
10
05
00

0 1 2 3

Number of Ice Cubes

Conclusions:

____

____

Group # _______ Date _______________ Names _______________ _______________

The Goldfish Experiment Page 2

_______________ _______________

Directions: Use the goldfish experiment sheet and other lab sheets to answer the questions.

Part II Questions:

1. What is a controlled experiment? _______________

2. Is the goldfish experiment a controlled experiment? Why or why not? _______________

3. What was the control and the variable in this experiment? _______________

4. List the controls in the goldfish experiment. _______________

5. Describe the variables in the goldfish experiment. _______________

Group # _______ Date ________________ Names ________________ ________________

________________ ________________

The Goldfish Experiment Page 3

6. Rewrite the problem in the goldfish experiment so it mentions breathing.

7. Rewrite your hypothesis in the goldfish experiment so it mentions the ice.

8. Write a brief lab report for the goldfish experiment. Mention the problem, the variables, and the results. Predict what would have happened if you had put another ice cube into the beaker. Also briefly tell what the experiment was about.

The Scientific Method Test

Names ____________________ ____________________

____________________ ____________________

Directions: Write the best answer in the space provided.

Part I Matching: Match the choices to the statements. Write the letter of the best answer on the line. Use each letter once.

	Statements	Choices
______	1. The orderly plan that a scientist uses.	A. Controlled experiment
______	2. A clear statement of what is to be done.	B. Controls
______	3. The form of the problem.	C. Experiment
______	4. A possible answer to the problem.	D. Hypothesis
______	5. One way to test the hypothesis.	E. Question
______	6. An experiment with many trials, many controls, and one variable.	F. Problem
______	7. The number of times an experiment is done.	G. Scientific method
______	8. The things in an experiment that do not change.	H. Collect information
______	9. The thing in the experiment that does change.	I. Trials
______	10. To read books and articles on a topic.	J. Variable

Part II The Laboratory: Put a checkmark in the space next to the things that should be in a laboratory. Leave the space next to the things that should not be in laboratory blank.

______ 11. Computers

______ 12. Food

______ 13. Test tubes

______ 14. Chemicals

______ 15. First-aid kit

______ 16. Books

______ 17. Magazines

______ 18. Live animals

______ 19. Drinks

______ 20. Beakers

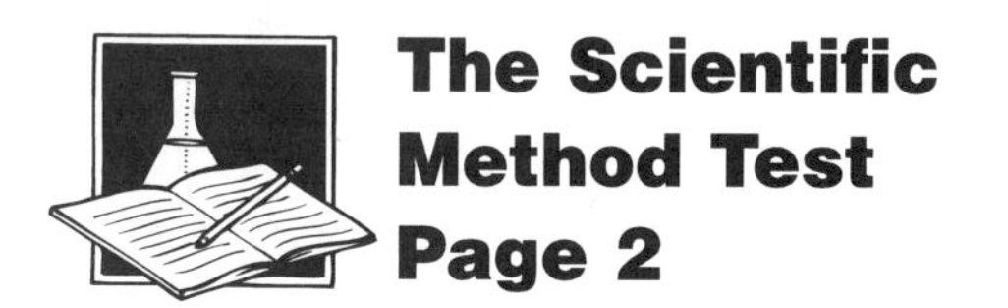

The Scientific Method Test Page 2

Name______________________________

Date ____________________ Hour__________

Part III The Scientific Method: Put the following steps of the scientific method in order. Write a 1 in the space next to the first thing to happen, then number the next steps 2, 3, 4, and 5 in order.

______ 21. Communicate the results

______ 22. Collect information

______ 23. Define the problem

______ 24. Form a hypothesis

______ 25. Test the hypothesis

Part IV Read the Story: Read the story carefully, then answer the multiple-choice questions on test page 3. Write the letter of the best answer in the space provided.

Maria's Mold

Maria is a middle-school student. One morning before school she went to get a glass of orange juice. When she got the pitcher out of the refrigerator, she noticed a blue mold growing in the juice. She poured the orange juice down the drain, washed the pitcher, and made a fresh pitcher of orange juice.

All day in school Maria thought about the mold. She wondered how it got there. She wondered how old the orange juice was. She wondered why it grew in the refrigerator. She asked her teacher questions about mold and read the chapter in her science book on mold.

When she got home she thought of an experiment that would answer some of her questions. Maria poured two identical cups of orange juice from the pitcher. She put one cup back in the refrigerator. She left the other cup on the kitchen counter. She asked her family not to drink or dump the cups of orange juice.

After three days she noticed mold growing on the orange juice left on the counter. The juice in the refrigerator was still good. She learned that the cold temperature in the refrigerator slowed down the growth of the mold.

The Scientific Method Test Page 3

Name________________________________

Date ____________________ Hour____________

______ 26. What is the best wording for the problem that Maria was trying to solve?

A. How does temperature affect the growth of mold?
B. Mold grows faster in a warm temperature.
C. What brand of orange juice tastes best?
D. What is the best place to grow mold?

______ 27. How did Maria collect information about the problem?

A. She asked her friends.
B. She asked her teacher and read her science book.
C. She called the orange juice company in Florida.
D. She read library books about mold.

______ 28. What is the best wording for Maria's hypothesis?

A. Cold temperatures slow down mold growth.
B. Mold grows from spores in the air.
C. Moldy orange juice doesn't taste very good.
D. Orange juice gets moldy faster than bread.

______ 29. How did Maria test her hypothesis?

A. She asked her teacher about mold and she read about mold in her science book.
B. She left a cup of orange juice in a place where there was lots of mold.
C. She drank moldy orange juice and timed how long it took her to get sick.
D. She put one cup of orange juice in a cold place and one cup in a warm place.

______ 30. In the story it does not mention how Maria communicated her results. What would be the best way for her to communicate her results?

A. Call the closest research laboratory and tell them about her findings.
B. Tell her friends about the experiment.
C. Tell the manager of the grocery store where she bought the orange juice.
D. Write a report and send it to the orange juice company in Florida.

______ 31. Maria controlled all of the following except:

A. Amount of light
B. Amount of orange juice
C. Brand of orange juice
D. Temperature

Chapter 3
Measurement

Teacher's Guide to Measurement

Scientific Measures Pamphlet (pages 63 and 64)

Have students use this as a reference for pages 66 to 68.

Scientific Measures Transparency Master (page 65)

Supplies Needed

Metric ruler
Meter stick
Triple beam balance
Graduated cylinder (any size)
Celsius thermometer
Stopwatch

Use this overhead master to introduce the measurement unit. Explain to the students that scientific measures today are done in the metric system. Also explain that most nations use the metric system. It's a good idea to show students items like soda cans and food boxes that have metric measures on them. Show students a balance, a meter stick, a graduated cylinder, and a Celsius thermometer. This is also a good time to introduce how the instruments are used. Talk about how easy it is to convert from one metric unit to another metric unit. Tell students that they need to forget about inches, quarts, pounds, and degrees Fahrenheit in science because in science only the metric system is used.

Metric Measures Study Guide (page 66)

This page should be done using the Scientific Measures Pamphlet as a reference. It is designed to be done individually but also works for groups. It's also helpful to have the transparency master on page 65 in view while students complete this page.

Metric Measures Predictions (page 67)

This page helps students apply the metric system to familiar objects. This page is best done individually. Students can use the Scientific Measures Pamphlet as a reference.

Metric Pictures Predictions (page 68)

These pictures are designed to help students visualize concepts such as time, temperature, distance, volume, and mass. Allow students to use the Scientific Measures Pamphlet to identify the pictures. This activity works well individually or in groups.

The Balance–Hands-on (page 69)

This activity is designed for students to learn to read the balance before they actually use one in an experiment. When students use a balance for the first time, they often cannot make it balance and correctly measure the mass of an object. The difficulty comes when students try to read the mass. For an example of this, see problem #1. Students may give the answer as 200304 grams. After completing this sheet students will understand that the correct mass is 234 g. It may help students to correctly interpret the balance if they have a balance on their table. Students should complete this page before they do a lab in which they are required to use a balance.

The Ruler–Hands-on (page 70)

Students can measure length, width, or height, but when it comes to recording mm and cm they have a difficult time. This activity will help students read the hash marks on the ruler. It may help students if they have a ruler at the table when they do this page. Students should complete this page before they do a lab in which they need to measure distance.

The Graduated Cylinder–Hands-on (page 71)

This page is designed to give students practice in reading a graduated cylinder. Students should complete this page before they are asked to measure volume in a lab. Make graduated cylinders available for student reference while completing this page. This is an excellent time to introduce the concept of a meniscus (the curve at the top of a liquid in a graduated cylinder). Teach students to read the bottom of a meniscus. For example, students often answer problem #1 as 22 mL. The correct answer is 20 mL because the bottom of the meniscus rests on the 20-mL mark.

The Thermometer–Hands-On (page 72)

The thermometer can be a very frustrating instrument for students to read because it involves the use of negative numbers. Students may have a hard time recording what they see on a thermometer. You may need to do a mini-unit on integers if students have had no exposure to negative numbers. This assignment helps students interpret what they see on a thermometer. Thermometers should be available for students to use as a reference for this activity. Students should complete this page before they do a lab in which they are required to use a thermometer.

The Metric Steps Transparency Master (page 73)

These steps are a tool students can use to help with converting within the metric system. For every step that you move down in the conversion, you move one decimal place to the right. For every decimal place that you move up in the conversion, you move one decimal place to the left. Metric System Conversions on page 75 presents problems for student practice.

Decimal Dos Transparency Master (page 74)

Students may know how to add, subtract, multiply, and divide decimals, but they may not know how to move decimals correctly. This transparency master presents four rules that teach students about decimals. Practice doing metric conversion problems with the class using these rules.

Metric System Conversions (page 75)

The problems in this activity are divided into three sets. Set A is designed to be done in class as a demonstration using the transparency masters on pages 73 and 74.

Set A, Problem #1

Here's how to do this problem using the metric steps.

1. Look at the problem and the steps. Notice that you are going from kg to g.
2. This conversion involves going down three steps.
3. Going down three steps tells us to go to the right three places.
4. Therefore, 4 kg = 4000 g.

Set B problems are designed to be done individually and then checked in a discussion format. Have students demonstrate how they arrived at their answers. Let them use the steps as a reference. Make sure every student has the correct answers to Sets A and B before the homework in Set C is assigned.

Set C is designed as homework. You may want to provide copies of pages 73 and 74 for students to use at home as a reference. Check Set C in class. Make sure every student has the right answers. If students had any wrong answers, make sure they know what they did wrong.

Metric System Conversions Quiz (page 76)

Pages 73-76 can be very frustrating for students. Remember, this unit is designed to expose them to metric conversions, and every student may not fully understand how to convert from one metric unit to another. The scores may very well be low on this quiz. To have a better understanding of this concept students may need more practice. Hand-check this quiz to learn which students understand conversions.

Do not mix the lessons on pages 73-76 with lessons on conversions between the metric units and standard units. Cover that at another time.

Metric System Distance Lab (pages 77 and 78)

Supplies Needed

Metric rulers, with mm and cm (8)
Meter sticks, with mm and cm (8)
Calculators, optional (8)
Nails, any size (8)
Paper clips (8)
Plastic beakers (8)
Coffee cans (8)
Plastic drinking straws (8)
Keys (8)
Quarters (8)
Books (8)

In this lab students are asked to measure the length, width, and height of several lines and objects. The best way for students to understand the units and instruments that scientists use to measure distance is to actually have students use the same units and instruments.

This lab is divided into two parts. In the first part, students estimate and measure the length of five line segments, some in mm and some in cm. They then calculate the difference between the estimated length and the actual length. Students should also calculate the total differences.

In Part II, students estimate and measure the length, width, or height of several objects. They are again asked to calculate the differences and total differences.

This lab is best done in cooperative groups and may take more than one class period. This and the next can be difficult to grade, especially with the estimates. A grading sheet has been provided for your convenience to use with both labs. Whether the grading sheet is used or not, it is imperative that great leeway be given on the estimates and very little (if any) leeway given on the actual measurements.

Metric System Volume Lab (pages 79 and 80)

Supplies Needed

100-mL graduated cylinders (8)
250-mL graduated cylinders (8)
Saturated salt solution (200 mL)
Rock salt crystals (40 crystals)
5 different towels (8 of each)

The lab provides an opportunity for students to practice using graduated cylinders, make estimations, make predictions, test the absorbency of towels, and calculate differences.

The lab is divided into two parts. In the first part, students need a saturated salt solution for the experiment. This solution is best made a day in advance, one or two liters at a time. Dissolve as much salt in the water as you can. A few undissolved grains may remain at the bottom of the mixing container. The rock salt crystals should not dissolve when students add them to the solution. Remember, the rock-salt crystals are used only to displace the water. Students can use calculators to do the addition and subtraction.

In Part II, students examine five towels, predict the rank in order how much each towel can absorb and hold, and estimate and measure how much water each towel can hold. Either paper towels, cloth towels, or a combination of both kinds will work for this lab. It's best to have each towel dry at the beginning of the lab. When students measure absorbency they won't be able to squeeze all the water out.

Metric System Mass Lab (page 82)

Supplies Needed

- Triple beam balance (8)
- Nails (8)
- 50-mL beakers (8)
- Paper clips (8)
- Nickels (8)
- Raisins (8)
- Scissors (8)
- Sticks of gum (8)
- Small rocks (8)
- Empty soda cans (8)
- Golf balls (8)

This lab provides an opportunity for students to practice using a balance scale and estimating and weighing a variety of objects.

This lab is best done in cooperative groups and can be difficult to grade. A grading sheet has been provided for your convenience.

Metric System Temperature Lab (page 84)

Supplies Needed

- Celsius thermometers (8)
- 500-mL glass beakers (8)
- Ice (small bag)
- Heat source (8)
- Stopwatches (8)
- 250-mL graduated cylinders (8)

This lab offers students an opportunity to practice measuring time and temperature. It also provides students with an opportunity to make a bar graph of the data they collected. Students will cool down a glass of water to as close to 0°C as they can. This is done by putting ice into the water. After reading the thermometer students begin to slowly heat the ice water and measure the temperature every minute. Hot plates, torches, lamps, or candles can be used as the heat source. The students are also required to record and graph their data.

Metric System Measuring Test (page 85)

Supplies Needed

- Celsius thermometers (8)
- Beakers (8)
- Triple beam balances (8)
- 100 or 250-mL graduated cylinders (8)
- Ice (small bag)
- Nickels (8)
- Freezer bags (8)
- Metric rulers, with cm and mm (8)

This test can be administered individually (if there is enough equipment), or in cooperative groups. Groups are recommended. Students are required to make an actual measurement of temperature, volume, mass, and distance. This test measures whether students can use the units and instruments scientists use.

Multiple Intelligences Applications

Verbal/Linguistic–As the Metric Measures Study Guide and other activities are checked in class, students will have opportunities to verbalize their answers. The Metric Pictures Predictions activity allows students to explain their ideas to their groups. Students can be asked to demonstrate on the board how they got their answers in the Metric System Conversions activity. When going over the quiz and test, students can explain their answers.

Logical/Mathematical–Students can apply accepted units and definitions of the metric system to activities in this chapter. They then follow the logical steps in converting from one metric unit to another.

Visual/Spatial–Students can see and actually use the various measuring devices. Many of the activities provide students with hands-on practice using metric measuring instruments. The transparency and Metric Pictures Predictions activity provide other visuals for the class. As an extension to the Metric System Conversions activity, students could represent various digits and one student could represent a decimal. Students can move the "decimal student" to the right or left as they convert from one metric unit to another.

Body/Kinesthetic–Students can walk, run, or jump various distances and use the metric system to measure and record the distances. They can demonstrate how to use the metric measuring instruments to the class.

Musical/Rhythmic–Students can write songs about using some of the metric units or instruments mentioned in this chapter. The teacher can play the song, "Hokey Pokey" while students move decimals to the right or left in the Metric System Conversions activity. Students can change the words in the song to make them fit decimals and the metric system.

Interpersonal–Students have many opportunities to explain their answers to their groups. They can explain the steps they used to move the decimals in the Metric System Conversions activity. When quizzes and tests are checked in class, students have an opportunity to discuss their answers.

Intrapersonal–Students can keep a log of the mass, distance, volume, time, and/or temperature of the items they measure. They can keep a list of how to use the instruments and the methods they used to solve metric conversion problems. Students can keep a journal of what they did in the various labs and write an essay about how they feel about metric measurement.

Volume

Distance

Mass

Temperature

Time

Scientific Measures

The Metric System

The Metric System

President George Washington first proposed that the United States adopt the metric system as its system of measurement. He was voted down.

Today, in the United States, the metric system is mainly used only by scientists.

The metric system is easy to learn once you know the correct prefixes and suffixes. Then it becomes a matter of simply moving decimals to the right or left the correct number of places.

The Suffixes

A suffix is a part of a word that comes at the end of the word. Metric system suffixes include:

- *-meter* for distance
- *-liter* for liquid volume
- *-gram* for mass

The Prefixes

A prefix is a part of a word that comes at the beginning of a word. Some metric prefixes include:

- kilo- means 1000 times
- centi- means one-hundredth
- milli- means one-thousandth

Putting It All Together

When you take a prefix like *milli-* and a suffix like *-meter* and put them together, you get the word *millimeter,* which means one-thousandth of a meter. Sometimes the suffixes are used alone, with no prefix. Common metric measures that are used in science include:

millimeter	centimeter	meter
milligram	gram	kilogram
milliliter	liter	

The Instruments

You will be asked to make detailed and accurate measures. Most of your instruments use the metric system.

You will use:

- Ruler for distance
- Balance for mass
- Graduated cylinder for volume
- Thermometer for temperature
- Stopwatch for time

Distance

Distance is a measurement of something's length, width, or height. The smallest distances are measured in millimeters. Millimeters are so small they can be used for measuring the length of an ant egg or the radius of a hair.

Centimeters are used to measure things like the width of this paper or the height of a person.

Meters are used for measuring things like the length of the room or the height of a mountain.

Kilometers are used for measuring things like the distance between New York and Los Angeles.

We use many instruments to measure distance. They include:

odometers in cars
meter sticks
rulers

Mass

A balance is an instrument commonly used to measure mass. Mass is different than weight. Weight is measured on a scale and measures the pull of gravity on an object. Mass measures the amount of matter an item contains.

Milligrams measure the mass of very small things like a dose of medicine or vitamins in food. Grams are used to measure larger things like the mass of the mustard in a bottle. Kilograms measure large things like the mass of a student.

Volume

Graduated cylinders are used to measure the volume of liquids. Liquids change shape and assume the shape of the container.

Milliliters measure small amounts of liquids like a can of soda. Liters are used for larger things like gas in a car's gas tank or water in a pool.

Temperature

Temperature is used to measure how hot or cold something is. Temperature is measured with a thermometer. In the metric system, scientists use degrees Celsius. This is an easy unit to use because water freezes at 0 degrees Celsius and boils at 100 degrees Celsius.

Time

Scientists use seconds, minutes, hours, days, months, and years to measure time. They use instruments like stopwatches, clocks, and calendars.

Symbols and Abbreviations

In the metric system, use of symbols and abbreviations is encouraged. The metric system has its own symbols plan. This plan gives symbols to prefixes and suffixes, and combines them when prefix/suffix combinations are used. A few common symbols are:

meter (m)	millimeter (mm)
liter (L)	centimeter (cm)
gram (g)	kilometer (km)
milli (m)	milligram (mg)
centi (c)	kilogram (kg)
kilo (k)	milliliter (mL)

Decimals

Since the metric system is based on the number *10*, it uses decimals. Once you become familiar with decimals, this will be a very easy system of measurement to use.

Scientific Measures

Measure	Definition	Unit(s)	Symbol(s)	Instrument(s)
Distance	Length, width, or height	Meter	m	Ruler
Mass	The amount of stuff	Gram	g	Balance
Liquid volume	The amount of space it takes	Liter	L	Graduated Cylinder
Temperature	How hot or cold something is	Degrees Celsius	°C	Thermometer
Time	How long it takes	Seconds	s	Stopwatch

Ruler

Balance

Graduated Cylinder

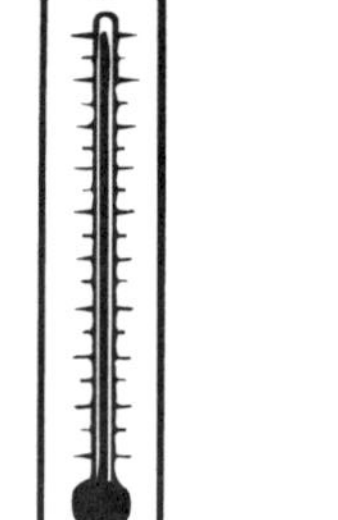

Thermometer

Stopwatch

Metric Measures Study Guide

Name______________________________

Date ________________ Hour__________

Directions: Use the Scientific Measures Pamphlet to answer the following questions.

Part I Instruments: Match the measure to the instrument. Use each choice once.

	Measure	Instrument
_____	1. Distance	A. Balance
_____	2. Mass	B. Graduated cylinder
_____	3. Temperature	C. Ruler
_____	4. Time	D. Stopwatch
_____	5. Volume	E. Thermometer

Part II Descriptions: Match the description to the measure. Use each choice once.

	Description	Measure
_____	6. How hot or cold something is	A. Distance
_____	7. How much material is in an object	B. Mass
_____	8. Amount of space something takes up	C. Temperature
_____	9. How long an event takes	D. Time
_____	10. How long, high, or wide something is	E. Volume

Part III Symbols: Write the correct symbol for each unit of measure.

_____ 11. Grams

_____ 12. Liters

_____ 13. Degrees Celsius

_____ 14. Meters

_____ 15. Centimeters

_____ 16. Milliliters

_____ 17. Millimeters

_____ 18. Seconds

_____ 19. Kilograms

_____ 20. Milligrams

Part IV Units: Match the measure to the unit. You may use choices more than once.

A. Distance B. Mass C. Temperature D. Time E. Volume

_____ 21. Meter

_____ 22. Gram

_____ 23. Second

_____ 24. Millimeter

_____ 25. Year

_____ 26. Milliliter

_____ 27. Minute

_____ 28. Millisecond

_____ 29. Liter

_____ 30. Degrees Celsius

_____ 31. Kilometer

_____ 32. Centimeter

_____ 33. Kilogram

_____ 34. Milligram

_____ 35. Month

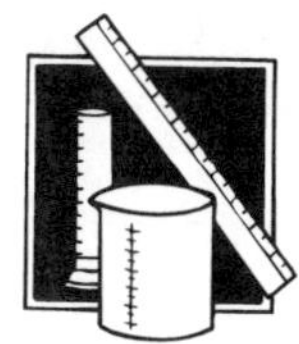

Metric Measures Predictions

Name____________________

Date ____________________ Hour____________

Directions: Predict what the best metric unit of volume, mass, distance, temperature, or time would be for the following situations. Write your answers on the lines.

1. The distance from Earth to the moon — 1. ____________
2. The average temperature of the moon — 2. ____________
3. The time it takes to get to the moon — 3. ____________
4. The mass of the moon — 4. ____________
5. The time it takes for sunlight to be reflected from the moon to Earth — 5. ____________
6. The amount of gas in the gas tank of a car — 6. ____________
7. The distance a car can travel on a tank of gas — 7. ____________
8. The mass of a car — 8. ____________
9. The volume of catsup you need for an order of fries — 9. ____________
10. The mass of sugar in a glass of chocolate milk — 10. ____________
11. The temperature of a cold glass of milk — 11. ____________
12. The volume of a glass of milk — 12. ____________
13. The mass of fat in an order of fries — 13. ____________
14. The mass of vitamin C in a glass of orange juice — 14. ____________
15. The distance a student can run in one minute — 15. ____________
16. The volume of water a student can drink after running one minute — 16. ____________
17. The mass of a student — 17. ____________
18. The height of a student — 18. ____________
19. The mass of a housefly — 19. ____________
20. The length of a housefly — 20. ____________
21. The distance from Chicago to Boston — 21. ____________
22. The height of Mt. Everest — 22. ____________
23. The distance from school to home — 23. ____________
24. The volume of water in a swimming pool — 24. ____________
25. The width of your classroom — 25. ____________
26. The volume of water in an average-sized water balloon — 26. ____________
27. The mass of your pen or pencil — 27. ____________
28. The length of your pen or pencil — 28. ____________
29. The height of your school — 29. ____________
30. The mass of your lunch — 30. ____________

Metric Pictures Predictions

Name____

Date ______________________ Hour______________

Directions: Study each picture. Tell whether the most obvious measure is time, distance, volume, mass, or temperature. Write the answers on the lines.

______________________ ______________________

______________________ ______________________

______________________ ______________________

______________________ ______________________

______________________ ______________________

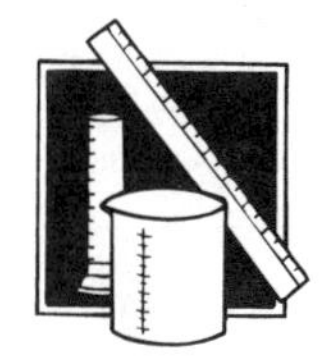

The Balance Hands-on

Name____________________

Date ______________ Hour__________

Directions: Determine what mass is shown on each triple beam balance. Write your answers on the lines.

1.

0		100		200		300				
0	10	20	30	40	50	60	70	80	90	100
0	1	2	3	4	5	6	7	8	9	10
0	0.1	0.2	0.3	0.4	0.5	0.6	0.7	0.8	0.9	1.0

__________ g

2.

0		100		200		300				
0	10	20	30	40	50	60	70	80	90	100
0	1	2	3	4	5	6	7	8	9	10
0	0.1	0.2	0.3	0.4	0.5	0.6	0.7	0.8	0.9	1.0

__________ g

3.

0		100		200		300				
0	10	20	30	40	50	60	70	80	90	100
0	1	2	3	4	5	6	7	8	9	10
0	0.1	0.2	0.3	0.4	0.5	0.6	0.7	0.8	0.9	1.0

__________ g

The Ruler Hands-on

Name____

Date ______________________ Hour______________

Directions: Determine what distance is shown on each ruler. Write your answers below the rulers. Be careful–some answers are in cm and others are in mm.

1. 1 2 3 4 5 6 7 8 9 10 11 12 13 14 15 16 17 18 19 20 21

___________ cm

2. 1 2 3 4 5 6 7 8 9 10 11 12 13 14 15 16 17 18 19 20 21

___________ cm

3. 1 2 3 4 5 6 7 8 9 10 11 12 13 14 15 16 17 18 19 20 21

___________ cm

4. 1 2 3 4 5 6 7 8 9 10 11 12 13 14 15 16 17 18 19 20 21

___________ mm

5. 1 2 3 4 5 6 7 8 9 10 11 12 13 14 15 16 17 18 19 20 21

___________ mm

The Graduated Cylinder Hands-on

Name____

Date ______________________ Hour______________

Directions: Determine what volume is shown on each graduated cylinder. Write your answers on the lines.

1.

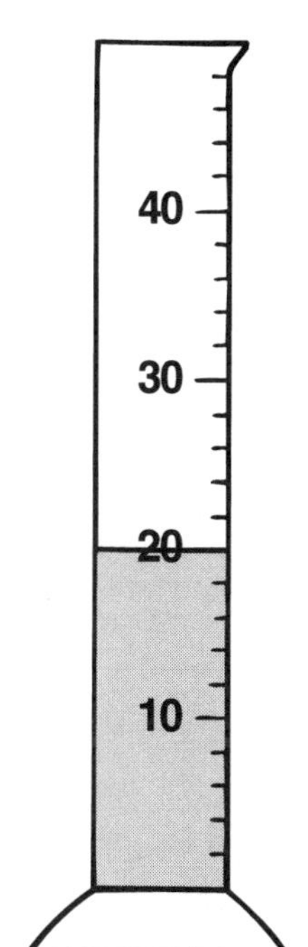

__________ mL

2.

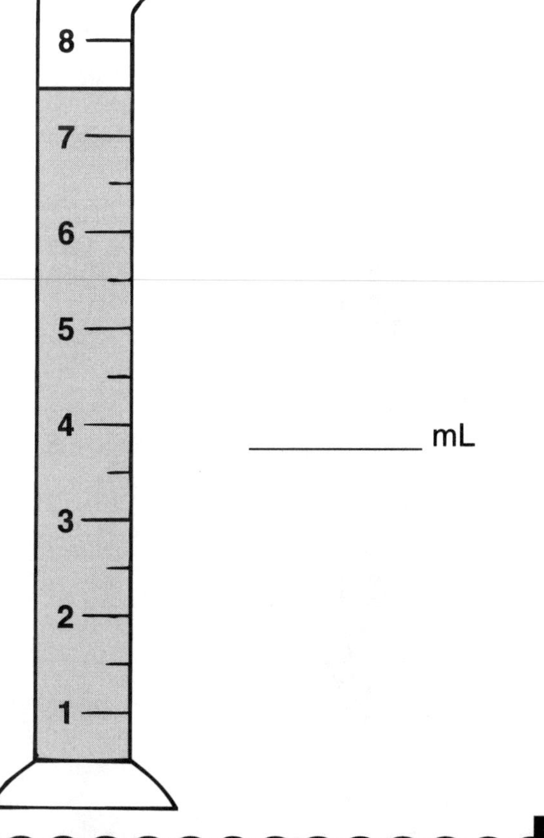

__________ mL

3.

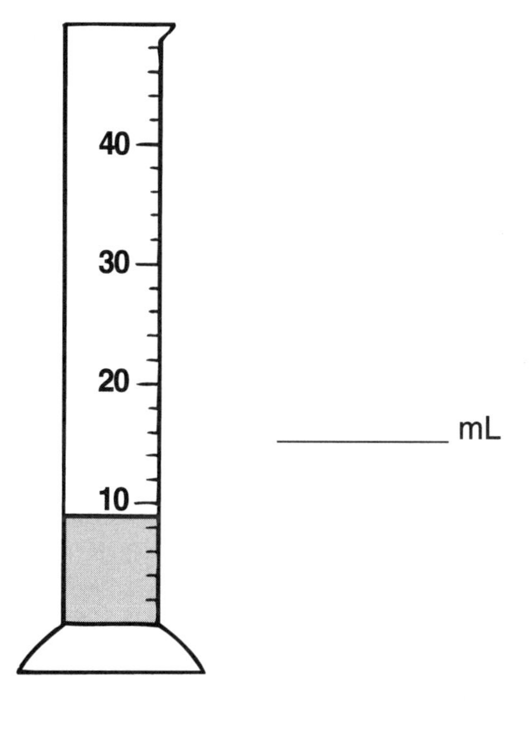

__________ mL

4.

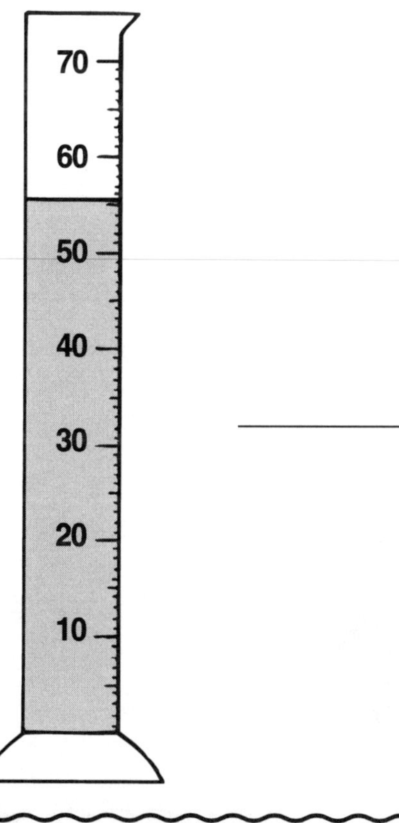

__________ mL

The Thermometer Hands-on

Name____

Date _______________________ Hour______________

Directions: Determine what temperature is shown on each thermometer. Write your answers on the lines.

1.

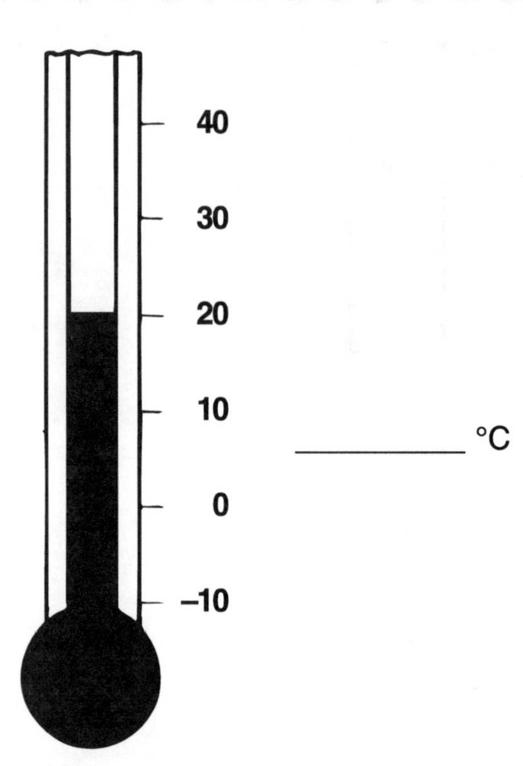

2.

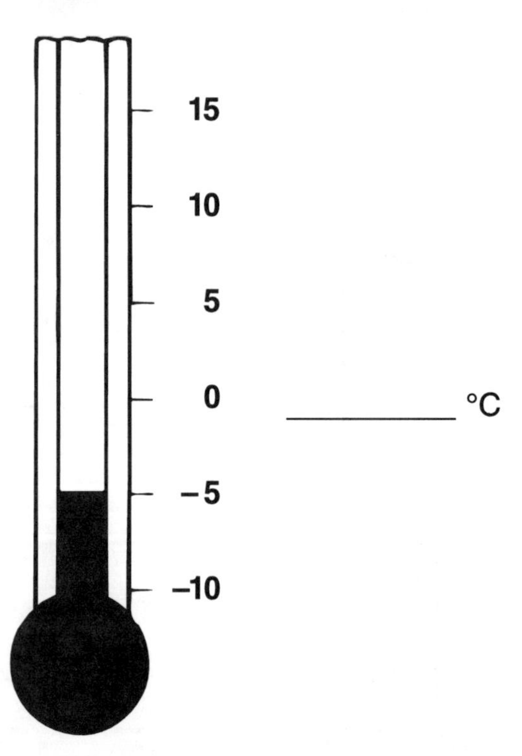

3.

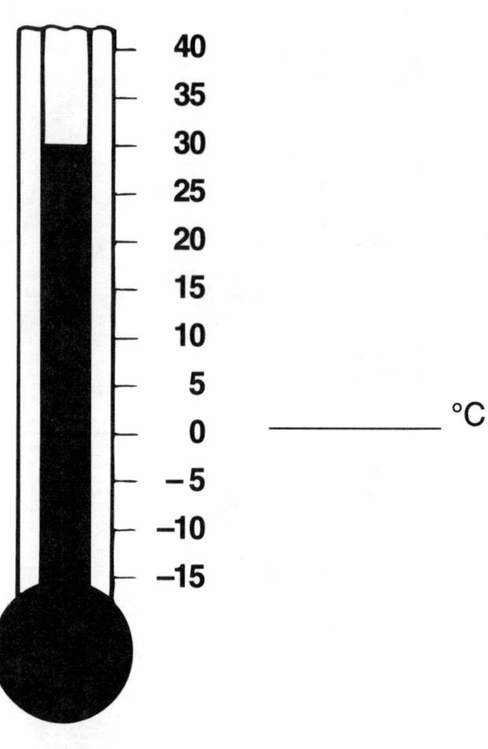

4.

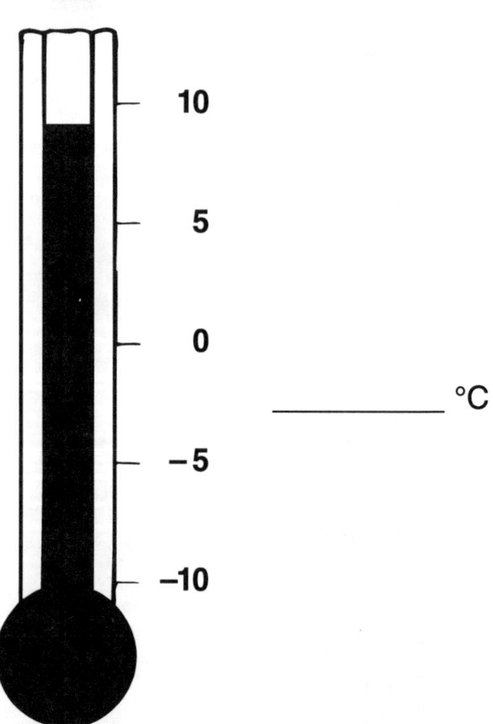

The Metric Steps

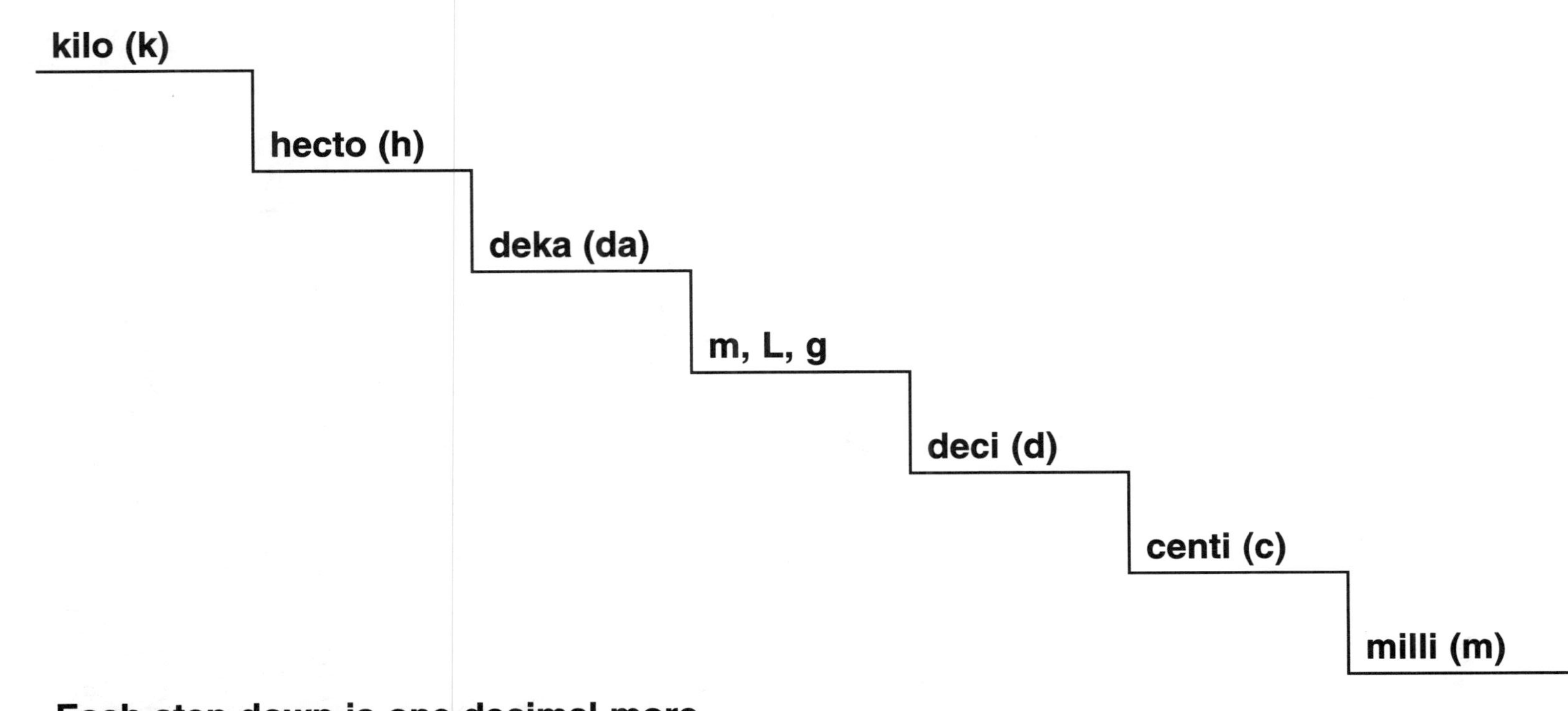

Each step down is one decimal more

Each step up is one decimal less

Decimal Dos

1. Every whole number has a decimal to the right of the last digit.

   15 = 15.0

2. You can put as many zeros as you want to the right of a decimal. This does not change the value of a number.

   15.0 = 15.00000

   0.38 = 0.380000

3. You can add as many zeros as you want to the left of a decimal. This does not change the value of a number.

   15.0 = 000015.0

   0.38 = 00000.38

4. In the metric system you can move the decimal one place to the right for every step down and one place to the left for every step up.

   32.0 cm = 320.0 mm

   32.0 cm = 0.32 m

Metric System Conversions

Name____

Date ____________________ Hour____________

Directions: Use the steps to move the decimals and make the conversions. Write your answers on the lines.

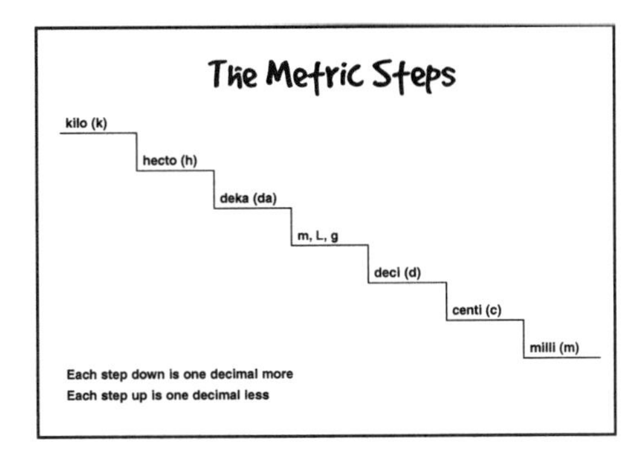

Set A: Practice

1. 4kg = ________ g
2. 3 km = ________ dam
3. 400 mL = ________ L
4. 3275 g = ________ kg

Set B: Individual

5. 100 m = ________ km
6. 1 g = ________ mg
7. 100 cm = ________ m
8. 1000 mL = ________ L
9. 75 kg = ________ g
10. 34 hm = ________ m
11. 600 mL = ________ L
12. 3275 g = ________ kg

Set C: Homework

13. 1 L = ________ mL
14. 493.6 cm = ________ m
15. 1.2 kg = ________ g
16. 0.0032 L = ________ mL
17. 0.186 km = ________ dm
18. 473.2 mL = ________ L
19. 723 m = ________ hm
20. 2041.3 g = ________ kg
21. 4.3 L = ________ mL
22. 0.043 kg = ________ g
23. 0.57 g = ________ mg
24. 1.91 L = ________ mL
25. 257 mm = ________ cm
26. 0.00094 kg ________ mg
27. 9342 mg = ________ g
28. 16 cm = ________ dm
29. 6900 mL = ________ L
30. 3407 mm = ________ m
31. 3407 mm = ________ cm
32. 3407 mm = ________ dm

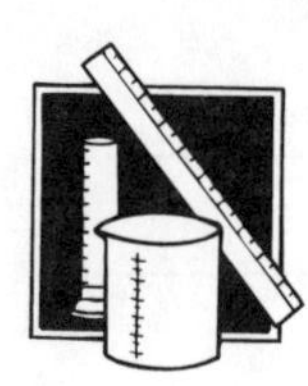

Metric System Conversions Quiz

Name____________________

Date ____________ Hour________

Directions: Use the steps to make the metric conversions. Write your answers on the lines.

1. 15 mg = ____________ g
2. 32.98 L = ____________ mL
3. 34 m = ____________ mm
4. 1268 m = ____________ km
5. 1.38 kg = ____________ g
6. 150 mm = ____________ m
7. 8.421 L = ____________ mL
8. 521 g = ____________ kg
9. 8.241 L = ____________ mL
10. 647 mL = ____________ L

kilo (k)
hecto (h)
deka (da)
m, L, g
deci (d)
centi (c)
milli (m)

Group # ______ Date ________________ Names ________________ ________________

________________ ________________

Metric System Distance Lab

Purpose: To practice using instruments and units scientists use to measure distance.

Part I Procedure:

1. Estimate the length of each line segment. Record the results in the Estimate column of the Data Chart.
2. Measure each line segment using a metric ruler. Record the results in the Actual column of the Data Chart.
3. Find the difference by subtracting. Record in the Difference column of the data chart.
4. Calculate your score by adding the total of the differences. Record at the bottom of the data chart marked score.

A. cm ____________________

B. cm ____________________________

C. cm ____

D. mm ___________________

E. mm ___________________________

Data Chart

	Estimate	Actual	Difference
A.	________	________	________
B.	________	________	________
C.	________	________	________
D.	________	________	________
E.	________	________	________
		Score (total of differences)	________

Group # ______ Date ______________ Names ______________ ______________

______________ ______________

Metric System Distance Lab Page 2

Part II Procedure:

1. Estimate the length, width, or height of each of the following objects. Record the results in the Estimate column of the Data Chart.
2. Use the best instrument and unit to measure the length, width, or height of each object. Record the results in the Actual column of the Data Chart.
3. Find the difference by subtracting. Record the results in the Difference column of the Data Chart.
4. Calculate your score by adding the total of the differences. Record the resulting score at the bottom of the Data Chart.

Data Chart

Object		Estimate	Actual	Difference
A. Length of the nail	**A.**	______	______	______
B. Width of this paper	**B.**	______	______	______
C. Height of the lab table	**C.**	______	______	______
D. Your height	**D.**	______	______	______
E. Width of the room	**E.**	______	______	______
F. Length of paper clip	**F.**	______	______	______
G. Length of your pen or pencil	**G.**	______	______	______
H. Height of coffee can	**H.**	______	______	______
I. Height of the beaker	**I.**	______	______	______
J. Length of the straw	**J.**	______	______	______
K. Length of the chalkboard	**K.**	______	______	______
L. Width of the door	**L.**	______	______	______
M. Length of the key	**M.**	______	______	______
N. Length of the diameter of a quarter	**N.**	______	______	______
O. Height of your book	**O.**	______	______	______
			Score (total of differences):	______

Group # ______ Date ________________ Names ________________ ________________

________________ ________________

Metric System Volume Lab

Purpose: To practice using the instruments and units scientists use to measure volume.

Part I Procedures:

1. Measure 20 mL of a saturated saltwater solution in a 100-mL graduated cylinder.
2. Predict how much a crystal of rock salt will raise the level of the water.
3. Record your prediction in the Estimate column of the Data Table.
4. Complete the entire Estimate column.
5. Add one crystal of rock salt to the 20 mL of saturated salt water.
6. Record the amount the level of the water rose in the Actual column of the Data Table.
7. Add another crystal of rock salt to the solution. Record the water level in the Actual column of the Data Table.
8. Repeat step seven until you have five crystals of rock salt in your graduated cylinder.
9. Find the differences by subtracting. Record the difference in the Difference column of the Data Table.
10. Calculate your score by adding the differences.

Data Table

Crystals of Rock Salt	Estimate	Actual	Difference
1			
2			
3			
4			
5			
		Score (total of differences)	

Group # ______ Date ______________ Names ______________ ______________

______________ ______________

Metric System Volume Lab Page 2

Part II Procedure:

1. Examine the five dry towels at your table.
2. Put them in order according to how much water you think each towel can hold in the table to the right. 1 holds the most and 5 holds the least.
3. Estimate, in mL, the amount of water each towel can hold.
4. Record the estimated amounts in the Estimate column in the Data Table below.
5. Saturate Towel A with tap water.
6. Squeeze all the water you can out of Towel A into the 250-mL graduated cylinder.
7. Record the amount in the Actual column in the Data Table.
8. Repeat steps 5-7 using Towels B, C, D, and E.
9. Find the differences by subtracting. Record the differences in the Difference column in the Data Table.
10. Calculate your score by adding the differences.

Rank Order	Towel
1	________
2	________
3	________
4	________
5	________

Data Table

Towel Letter	Estimate (mL)	Actual (mL)	Difference
A			
B			
C			
D			
E			
		Score (total of differences)	

Group # ______ Date ________________ Names ____________________ ____________________

____________________ ____________________

Grading Sheet

This assessment tool can be used for the Distance Lab and Volume Lab activities.

Part I

1. **Estimate** — 1 2 3 4 5 6 7 8 9 10
   How close were the estimates to the actual?

2. **Actual** — 1 2 3 4 5 6 7 8 9 10
   How close was the measuring?

3. **Subtractions and Additions** — 1 2 3 4 5 6 7 8 9 10
   How accurately were the subtractions and additions done?

4. **Score** — 1 2 3 4 5 6 7 8 9 10
   How close was the total of differences to 0?

Part II

5. **Objects** — 1 2 3 4 5 6 7 8 9 10
   Were the objects measured correctly?

6. **Estimates** — 1 2 3 4 5 6 7 8 9 10
   How close were the estimates to the actual?

7. **Subtractions and Additions** — 1 2 3 4 5 6 7 8 9 10
   How accurately were the subtractions and additions done?

8. **Score** — 1 2 3 4 5 6 7 8 9 10
   How close was the total of differences to 0?

Overall

9. **Care and Effort** — 1 2 3 4 5 6 7 8 9 10
   Was the lab carefully done?

10. **Measuring and Estimating** — 1 2 3 4 5 6 7 8 9 10
    Did the measuring and estimation get better as the lab progressed?

Total (100) __________

Group # _______ Date _________________ Names ______________________ ______________________

______________________ ______________________

Metric System Mass Lab

Purpose: To practice using the units and instruments scientists use to measure mass.

Procedures:

1. Examine the nail, beaker, paper clip, nickel, raisin, scissors, gum, rock, can, and golf ball.
2. Rank them in order from greatest to least mass. 1 is the greatest and 10 is the least.
3. Record the order on the chart to the right.
4. Use the balance to measure the mass of each object.
5. Record the mass of each object in the Data Table below.
6. Make a bar graph showing the mass of each object.

Rank Order	Object
1	
2	
3	
4	
5	
6	
7	
8	
9	
10	

Data Table

Object	Mass (g)	Object	Mass(g)
Nail		Scissors	
Beaker		Gum	
Clip		Rock	
Nickel		Can	
Raisin		Golf Ball	

Bar Graph

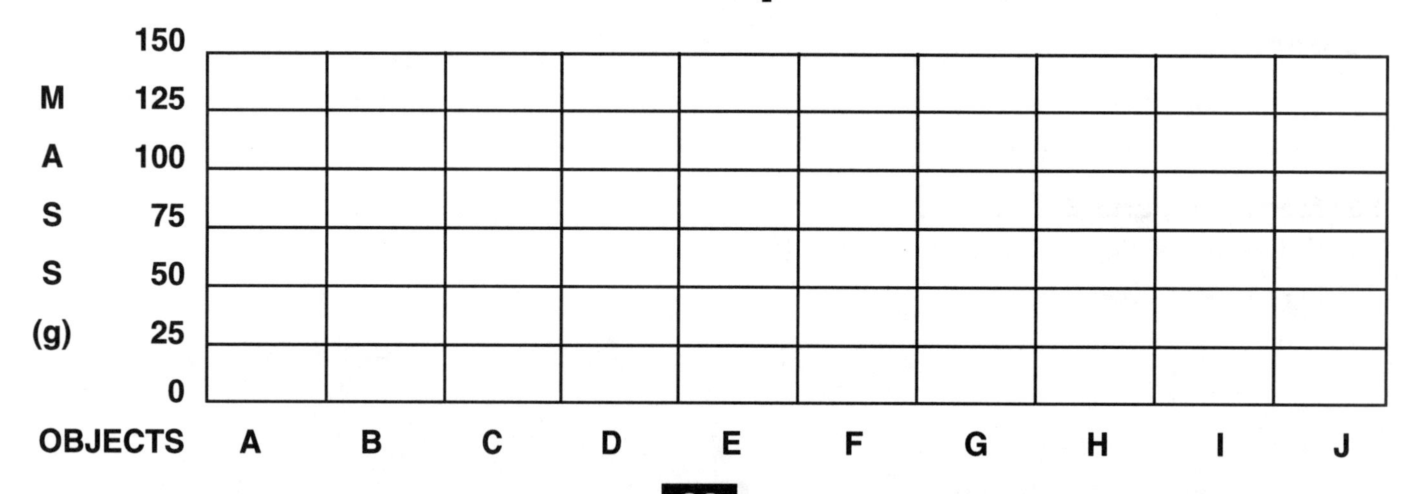

Group # ______ Date ________________ Names ________________ ________________

________________ ________________

Mass Lab Grading Sheet

1. Rank Order 1 2 3 4 5 6 7 8 9 10 11 12 13 14 15 16 17 18 19 20 21 22 23 24 25
How close was the estimated order to the actual order?

2. Data Table 1 2 3 4 5 6 7 8 9 10 11 12 13 14 15 16 17 18 19 20 21 22 23 24 25
How close was the measured mass to the actual mass?

3. Bar Graph 1 2 3 4 5 6 7 8 9 10 11 12 13 14 15 16 17 18 19 20 21 22 23 24 25
Is the bar graph planned out, neat, and accurate?

4. Overall 1 2 3 4 5 6 7 8 9 10 11 12 13 14 15 16 17 18 19 20 21 22 23 24 25
Are the results accurate? Was the paper done neatly? Was the lab completed in the amount of time allowed?

Total (100): __________

Group # ______ Date ________________ Names ________________ ________________

________________ ________________

The Metric System Temperature Lab

Purpose: To practice using the units and instruments scientists use to measure temperature.

Procedure:

1. Make a solution of ice water in the beaker using 250 mL of water and three to four ice cubes.
2. Put the thermometer into the ice water and watch the temperature decrease to between 0°C and 5°C. Record this temperature in the 0-minute section on the Data Table below.
3. Place the ice water with the thermometer on the heat source.
4. Begin heating, and start the stopwatch at the same time.
5. Record the temperature of the water every minute in the Data Table.
6. Stop the heat source and end the experiment when the temperature reaches 100°C or the time reads 10 minutes.
7. Make a bar graph of your data.

Data Table

Time (min)	Temperature (°C)	Time (min)	Temperature (°C)
0		6	
1		7	
2		8	
3		9	
4		10	
5			

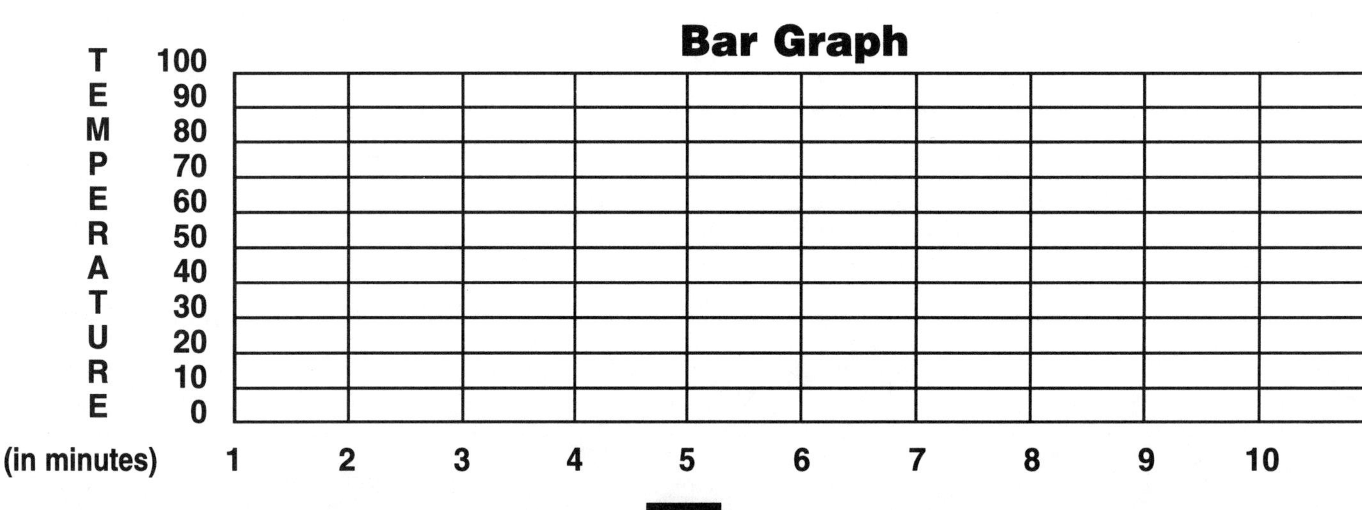

Group # ______ Date ________________ Names ________________ ________________

________________ ________________

Metric System Measuring Test

Directions: Use the Celsius thermometer, beaker, balance, graduated cylinder, ice, nickel, freezer bag with tap water, and ruler to make the following measurements. Write your answers. Use the proper symbols for the units in your answers.

1. Measure the temperature of the room. ________________
2. Measure the temperature of a beaker of ice water. ________________
3. Measure the volume of water in the freezer bag. ________________
4. Measure the mass of a nickel. ________________
5. Measure the mass of the dry beaker. ________________
6. Measure the mass of 50 mL of tap water. ________________
7. Measure the length of your paper in mm. ________________
8. Measure the width of your paper in mm. ________________
9. Measure the height of the beaker in mm. ________________

10.–12. Measure the length of the following lines in cm. Write your answers on the lines.

10. ____________

11. __

12. ____

Answer Key

Chapter 1 Laboratory Safety

Lab Safety Rules Pictures (Page 14)

Answers will vary. Accept any answer students can justify. Possible answers include:

Fire Extinguisher: Know where all the safety items are located. The fire extinguisher is a lab safety item.

Flame: Never reach over an open flame. Picture is an open flame.

Bandages: Know where all the safety items are located. The bandages are part of a first-aid kit.

Safety Glasses: Wear aprons and safety glasses when told to do so. Safety glasses are shown.

Eye: Never look directly into bright lights. They can damage your eyes.

Chocolate Bar: Never put anything in or near your mouth or eyes. Candy should not be eaten in the lab.

Apron: Wear aprons and safety glasses when told to do so. The apron in the picture will protect people and their clothing.

Match: Dispose of used chemicals and matches as your teacher instructs. The match in the picture needs to be completely out before it can be thrown away.

Teacher at Board: Listen carefully to the teacher's directions. It looks as if the teacher is giving directions.

Raised Hand: Wash your hands. It looks like a clean hand.

Smiling Girl: Tie back long hair.

Shower: Know where all the safety items are located. A safety shower is a safety item in the lab.

Burner and Flask: Never reach over an open flame. There is an open flame shown in the picture.

Man Pouring Chemical: Never leave chemical bottles open.

Safe School Labs Study Questions "New Sweatshirt" (Page 16)

1. The students were staining onion cells, observing them under the microscope, and drawing them.
2. Never leave chemical bottles open. Keep all unnecessary items off the lab table. Prepare properly. Listen carefully to the teacher's directions. (Students can have any two of these rules as long as they can justify their answers.)
3. Tom did not have to wear the new sweatshirt. Tom did not have to be drawing at the lab table. Tom could have worn an apron. (Any one of these answers will do, or any answer students can justify.)
4. Lafonzo could have put the top on the iodine bottle.
5. Mr. Washington might have noticed the top off the iodine bottle and instructed Lafonzo to replace it. He could have asked Tom to go to the locker room to change his shirt. (Answers will vary. Accept any answer students can justify.)
6. Accept any answer that represents the story.
7. Accept any well-written answer.
8. Accept any well-written answer.

Safe School Labs Study Questions
"Ms. Reitman's Lesson" (Page 17)

1. The class would evaporate sea water and collect and cool the water vapor.
2. Never force glass. Listen carefully to the teacher's directions.
3. Marla could have listened to the teacher and not forced glass.
4. Ms. Reitman did what she was supposed to do. She could have made sure Marla was listening and made sure that she didn't force glass. She could have had the class work in groups.
5. Other members of the class could have notified Ms. Reitman that Marla didn't know what to do. They could have helped Marla.
6. Any answer that relates to the story is acceptable.
7. Any positive answer that relates to the story is acceptable.
8. The accident was Marla's fault for not following the lab safety rules.

Safe School Labs Stories Quiz
(Pages 18 and 19)

Part I: 1. C; 2. E; 3. B; 4. F; 5. D; 6. I; 7. A; 8. G; 9. H.

Part II: The following items should be checked: 11, 13, 15, 16, 17, 18.

Part III: 21. 4; 22. 8; 23. 2; 24. 6; 25. 1; 26. 7; 27. 3; 28. 5.

Part IV: 29. 3; 30. 6; 31. 9; 32. 4; 33. 7; 34. 2; 35. 1; 36. 8; 37. 5.

Lab Safety Group Activity
(Pages 20-22)

Answers will vary. Any answers students can justify are acceptable. Here are some possible answers:

1. Problem: No safety glasses or aprons.
   Safe Practice: Lab table is clear.
2. Problem: Students are engaging in horseplay.
   Safe Practices: Students are wearing safety glasses; long hair is tied back.
3. Problems: The boy is not wearing his goggles properly. Girl is reaching over flame. Girl's long sleeves not rolled up.
   Safe Practices: Both students are wearing aprons. The girl is wearing her safety glasses.
4. Problem: The girl on the far left did not tie her hair back.
   Safe Practices: Students are using the instruments safely. All the students are wearing their safety glasses.
5. Problems: The girl on the right did not tie her hair back properly. Girl's long sleeves are not rolled up.
   Safe Practices: The lab table is neat. Students are using the instruments properly.
6. Problems: Students don't seem to know what to do. Boy on left is wearing goggles improperly.
   Safe Practices: The students have their aprons on properly.
7. Problem: The boy needs to dispose of the match properly.
   Safe Practice: The boy is using the instrument properly.
8. Problems: The boy has his hand near his mouth and is eating. Lab table is cluttered.
   Safe Practices: The boy is wearing safety glasses. The boy is wearing his apron.
9. Problem: The girl has not properly tied her hair back.
   Safe Practices: The girl has on safety glasses and is working on a neat lab table.
10. Problems: The boy on the left is not wearing safety glasses and has his hands near his eyes. The boy on the right is reaching over an open flame.
    Safe Practices: Both boys are wearing aprons. Boy on left has hair tied back. Boy on right is wearing safety glasses.

Laboratory Safety Test (Pages 24-26)

Part I: 1. C; 2. A; 3. D; 4. B; 5. B; 6. D; 7. A; 8. C; 9. A; 10. B; 11. D; 12. B;

Part II: 13. C; 14. D; 15. A; 16. C; 17. A; 18. D; 19. C; 20. B.

Chapter 2 The Scientific Method

Scientific Method Pictures (Page 35)

Answers will vary. Possible answers include:

Books:

Step: Collect Information

Reason: Scientists use books to collect information.

Mice:

Step: Test the Hypothesis

Reason: Scientists often use animals to test a hypothesis.

Question Mark:

Step: Define the Problem

Reason: All problems end with a question mark.

Milk:

Step: Test the Hypothesis

Reason: Scientists use products like white milk to test the hypothesis.

Burner and Flask:

Step: Test the Hypothesis

Reason: These instruments are often used in experiments.

Camera:

Step: Communicate the Results

Reason: Scientists use photographs to communicate results.

Plant:

Step: Test the Hypothesis

Reason: Sometimes scientists use plants to test a hypothesis.

Scientific Method Study Questions (Pages 36 and 37)

Answers will vary. Possible answer includes:

1. Books, computer, magazines, reports, plants, animals, chemicals, fire extinguisher, first-aid kit, collections.
2. Any problem not mentioned in the pamphlet will do.
3. Going to the library, using the Internet, reading magazines, books and reports specific to the problem the student has written.
4. Any statement that answers the question in #2 is acceptable.
5. Any experiment that tests the hypothesis in #4 is acceptable.
6. Accept any report that describes the results of the experiment in #5.
7. 1. Define the Problem
   2. Collect Information
   3. Form a Hypothesis
   4. Test the Hypothesis
   5. Communicate the Results

Be sure to stress to the students that good science often leads to more questions.

Scientific Method Mini-Posters (Page 38)

Accept any posters that follow the rubric on this page 39. Allow students to see the rubric before they do the mini-posters.

Scientific Method Homework (Page 40)

Student answers will vary on this homework assignment. Make sure students follow the scientific method.

Scientific Method Quiz (Page 41)

Part I: 1. D; 2. C; 3. B; 4. A; 5. E.

Part II: 6. A; 7. B; 8. A; 9. A; 10. B.

Part III: 11. D; 12. A; 13. B; 14. E; 15. C.

Bean Plant Experiment
(Pages 42-43)

Answers on this lab will vary.

10. Accept any prediction that relates to the bean plants. Here are some possible predictions.
    Plant A–This plant will continue to grow in a normal fashion.
    Plant B–This plant will curve and grow up.
    Plant C–This plant will curve and grow all the way around.
12. Students should observe that the stems of the plants will grow away from gravity if they have done the experiment correctly.
13. Students should observe that the roots will grow toward the center of gravity if they have done the experiment correctly.
14. Any laboratory report that follows the scientific method and the results of the experiment will do.

Bean Plant Experiment Write-up
(Page 44)

Answers will vary on this write-up. Possible answer include:

1. How do bean plants respond to changes in gravity?
2. Students should have the same answer as they did in #10 of the lab sheet (page 42).
3. Yes, because two of the plants were changed and one was not changed.
4. The variables were Plants B and C.
5. Plant A, amount of water, type of soil, amount of sunlight, and temperature.
6. The unchanged plant (Plant A) should be drawn here.

Bean Plant Experiment Summary Sheet
(Pages 45-47)

1. B; 2. E; 3. D; 4. C; 5. B; 6. C; 7. D; 8. A; 9. B; 10. B; 11. D; 12. D; 13. A; 14. C; 15. D.

Measuring Air Temperature Lab
(Pages 48-50)

Answers in this lab will vary. Possible answer include:

Data Table 1: The temperature should increase until it returns to room temperature. It should take about three to five minutes.

Data Table 2: The temperature should increase until it gets back to room temperature. It should take longer for the temperature to get to room temperature. It should take about four to six minutes.

Graph: This graph should show two different-colored lines. The first color should show the shaken thermometer; the curve should show a rapid rise to room temperature. The second color should show the still thermometer and a somewhat slower rise to room temperature.

Questions:

1. It wasn't shaken. There was a smaller amount of friction against the thermometer to help raise the temperature.
2. Faster. The liquid in the glass has less distance to travel.
3. Will a thermometer react faster if it is shaken rather than held still?
4. The shaken thermometer will react faster.
   The still thermometer will react faster.
   There is no difference in the reaction time of the two types of thermometers.
5. The shaken thermometer reacted faster.
6. The way the thermometer was held.
7. This lab report should follow the scientific method and indicate the results of the experiment.

The Goldfish Experiment
(Page 51-53)

Part I: Answers will vary in this lab. Possible answer include:

Hypothesis: There are four possible hypotheses for this experiment:

1. As temperature decreases, the goldfish

will breathe faster.

2. As temperature decreases, the goldfish will breathe slower.
3. As temperature decreases, the respiration of the goldfish will not change.
4. As temperature decreases, the goldfish will die.

Table 1–Data: This table should show a steady decrease in the respiration of the goldfish.

Graph 1–Goldfish Respirations: This should be a line graph showing the decrease in goldfish respirations.

Conclusions: As temperature decreased, goldfish respirations decreased.

Part II:

1. An experiment that has many controls, many trials, and one variable.
2. Yes. If class data is used there were many trials, many controls, and one variable.
3. The control was the individual goldfish. The variable was the change in temperature.
4. The individual goldfish, type of water, time, and amount of food.
5. The only thing students changed was the temperature (amount of ice cubes). This caused the goldfish respirations to change.
6. How will goldfish change their breathing when the temperature of the water changes?
7. Goldfish respirations will decrease as the number of ice cubes increase.
8. Here students should follow the scientific method and indicate the results of their experiment.

The Scientific Method Test (Pages 54-56)

Part I: 1. G; 2. F; 3. E; 4. D; 5. C; 6. A; 7. I; 8. B; 9. J; 10. H.

Part II: The following items should be checked: 11, 13, 14, 15, 16, 17, 18, 20.

Part III: 21. 5; 22. 2; 23. 1; 24. 3; 25. 4

Part IV: 26. A; 27. B; 28. A; 29. D; 30. D; 31. D;

Chapter 3 Measurement

Metric Measures Study Guide (Page 66)

Part I: 1. C; 2. A; 3. E; 4. D; 5. B.

Part II: 6. C; 7. B; 8. E; 9. D; 10. A.

Part III:.11. g; 12. L; 13. °C; 14. m; 15. cm; 16. mL; 17. mm; 18. s; 19. kg; 20. mg.

Part IV: 21. A; 22. B; 23. D; 24. A; 25. D; 26. E; 27. D; 28. D; 29. E; 30. C; 31. A; 32. A; 33. B; 34. B; 35. D.

Metric Measures Predictions (Page 67)

1. Kilometers
2. Degrees Celsius
3. Days or weeks
4. Kilograms
5. Minutes
6. Liters
7. Kilometers
8. Kilograms
9. Milliliters
10. Milligrams
11. Degrees Celsius
12. Milliliters
13. Milligrams
14. Milligrams
15. Meters
16. Milliliters
17. Kilograms
18. Centimeters or meters
19. Grams or Milligrams
20. Centimeters or Millimeters
21. Kilometers
22. Kilometers or meters
23. Meters or kilometers
24. Liters
25. Meters

26. Milliliters
27. Grams
28. Centimeters
29. Meters
30. Kilograms

Metric Pictures Predictions (Page 68)

Thermometer–Temperature
Ruler–Distance
Water bottle–Volume
Hiker–Distance
Weight Lifter–Mass
Flasks and graduated cylinder–Volume
Burner and flask–Temperature or volume
Stopwatch–Time
Elephant–Mass
Hurtle–Time or distance

The Balance–Hands-on (Page 69)

1. 234.0g; 2. 80.6g; 3. 9.25g.

The Ruler Hands-on (Page 70)

1. 14 cm; 2. 6.5 cm; 3. 16.8 cm;
4. 30 mm; 5. 7 mm.

The Graduated Cylinder–Hands-on (Page 71)

1. 20 mL; 2. 7.5 mL; 3. 9 mL;
4. 55.5 mL.

The Thermometer–Hands-on (Page 72)

1. 20°C; 2. - 5°C; 3. 30°C; 4. 9°C.

Metric System Conversions (Page 75)

1. 4000 g; 2. 300 dam; 3. 0.4 L; 4. 3.275 kg;
5. 0.1 km; 6. 1000 mg; 7. 1 m; 8. 1 L.; 9. 75000 g; 10. 3400 m; 11. 0.6 L; 12. 3.275 kg. 13. 1000 mL; 14. 4.936 m; 15. 1200 g;
16. 3.2 mL; 17. 1860 dm; 18. 0.4732 L;
19. 7.23 hm; 20. 2.0413 kg; 21. 4300 mL;
22. 43 g; 23. 570 mg; 24. 1910 mL;
25. 25.7 cm; 26. 940 mg; 27. 9.342 g;
28. 1.6 dm; 29. 6.9 L; 30. 3.407 m;
31. 340.7 cm; 32. 34.07 dm.

Metric System Conversions Quiz (Page 75)

1. 0.015 g; 2. 32980 mL; 3. 34000 mm;
4. 1.268 km; 5. 1380 g; 6. 0.15 m;
7. 8421 mL; 8. 0.521 kg; 9. 8421 mL;
10. 0.647 L.

The Metric System Distance Lab (Page 77-78)

Part I:

The Estimate and Difference columns will vary. Here are the answers for the Actual column:

A. 4 cm; B. 5.5 cm; C. 0.9 cm; D. 38 mm;
E. 52 mm.

Part II:

The Estimate and Difference columns will vary. Most of the Actual column answers will vary. Here are a couple answers for the Actual column:

B. 21.6 cm
N. 2.4 cm

The Metric System Volume Lab (Pages 79-80)

Data Table: Answers will vary depending on the size of the salt crystal.

Towel Rank Order: Answers will vary depending on the types of towels used.

Data Table: Answers will vary depending on the kinds of towels used.

Metric System Mass Lab (Page 82)

Rank Order Table: Answers will vary depending on the objects used.

Data Table: Most of the answers will vary depending on the types of objects used. Here are a couple that do not vary much:

Nickel = 5 g
Raisin = 1 g

Bar Graph: The bar graph should reflect the data in the data table.

The Metric System Temperature Lab (Page 84)

Data Table: Answers will vary. They should show a steady increase in temperature. At 0 minutes the temperature should be as close to 0°C as possible. Students should attempt to get the temperature to 100 degrees Celsius in 10 minutes.

Bar Graph: The bar graph should reflect the data in the data table. It should show a steady increase in temperature.

The Metric System Measuring Test (Page 85)

1. Anything between 19 and 24°C.
2. Anything between 0 and 5°C.
3. It is suggested using 250 mL of water.
4. 5 g
5. Answers will vary depending on the size and type of beaker.
6. It should be around 50 g.
7. 280 mm
8. 216 mm
9. Answers will vary depending on the type and size of beaker.
10. 2.4 cm
11. 0.6 cm
12. 8.7 cm